Applied Mathematical Sciences

EDITORIAL STATEMENT

The mathematization of all sciences, the fading of traditional scientific bounda-
ries, the impact of computer technology, the growing importance of mathematical-
computer modelling and the necessity of scientific planning all create the need both
in education and research for books that are introductory to and abreast of these
developments.

The purpose of this series is to provide such books, suitable for the user of
mathematics, the mathematician interested in applications, and the student scientist.
In particular, this series will provide an outlet for material less formally presented and
more anticipatory of needs than finished texts or monographs, yet of immediate interest because of the novelty of its treatment of an application or of mathematics
being applied or lying close to applications.

The aim of the series is, through rapid publication in an attractive but inexpen-
sive format, to make material of current interest widely accessible. This implies the
absence of excessive generality and abstraction, and unrealistic idealization, but with
quality of exposition as a goal.

Many of the books will originate out of and will stimulate the development of
new undergraduate and graduate courses in the applications of mathematics. Some
of the books will present introductions to new areas of research, new applications
and act as signposts for new directions in the mathematical sciences. This series will
often serve as an intermediate stage of the publication of material which, through
exposure here, will be further developed and refined. These will appear in conven-
tional format and in hard cover.

MANUSCRIPTS

The Editors welcome all inquiries regarding the submission of manuscripts for
the series. Final preparation of all manuscripts will take place in the editorial offices
of the series in the Division of Applied Mathematics, Brown University, Providence,
Rhode Island.

SPRINGER-VERLAG NEW YORK INC., 175 Fifth Avenue, New York, N.Y. 10010

Applied Mathematical Sciences | Volume 65

Applied Mathematical Sciences

Richard H. Rand
Dieter Armbruster

Perturbation Methods, Bifurcation Theory and Computer Algebra

With 10 Illustrations

Springer-Verlag
New York Berlin Heidelberg
London Paris Tokyo

Richard H. Rand
Department of Theoretical
  & Applied Mechanics
Cornell University
Ithaca, New York 14853
USA

Dieter Armbruster
Institut für Informations-
verarbeitung
Universität Tübingen 74 Tübingen 1
FRG

AMS Subject Classification: 34A34/68C20/34CXX/35B32/34D10/70KXX/70H05/70H15

Library of Congress Cataloging in Publication Data
Rand, R. H. (Richard H.)
  Perturbation methods, bifurcation theory, and
computer algebra.
  (Applied mathematical sciences ; v. 65)
  Bibliography: p.
  Includes index.
  1. Perturbation (Mathematics)   2. Bifurcation
theory.   3. Algebra—Data processing.   4. MACSYMA
(Computer system)   I. Armbruster, Dieter.   II. Title.
III. Series: Applied mathematical sciences (Springer-
Verlag New York Inc.) ; v. 65.
QA1.A647   vol. 65   510 s   87-16703
[QA871]   [515.3′53]

Text prepared by the authors in camera-ready form.
Printed and bound by R.R. Donnelley & Sons, Harrisonburg, Virginia.
Printed in the United States of America.

9 8 7 6 5 4 3 2 1

ISBN 0-387-96589-0 Springer-Verlag New York Berlin Heidelberg
ISBN 3-540-96589-0 Springer-Verlag Berlin Heidelberg New York

Contents

Preface

Our purpose in writing this book is to provide computer algebra programs which implement a number of popular perturbation methods. For each perturbation method, we present an introduction to the method, a couple of example problems, sample runs of the computer algebra programs and complete program listings.

In addition, we include examples of various elementary bifurcations, such as folds, pitchforks and Hopf bifurcations. These arise in the example problems. Specifically, we treat Hopf bifurcations in autonomous nonlinear systems via Lindstedt's method, the construction of center manifolds for simple, degenerate and nilpotent bifurcations in ordinary differential equations, the determination of normal forms for Hopf bifurcations and Takens-Bogdanov bifurcations, and averaging for autonomous and nonautonomous systems. Further, we use Lie transforms to determine normal forms in Hamiltonian systems. Bifurcation in partial differential equations, such as reaction diffusion equations or the Bernard convection problem, are treated via Liapunov-Schmidt reduction.

Moreover, we offer comparisons of the various methods. We compare averaging with normal forms, Liapunov-Schmidt reduction with center manifold reduction, Lindstedt's method with normal form calculations, and so on. To help in making the comparisons we frequently treat the same problem by two or more methods. E.g., we derive the Hopf bifurcation formula both by Lindstedt's method as well as via normal forms.

Our motivation for applying computer algebra to perturbation problems comes from the nature of the computations involved in these kinds of problems.

The massive algebra usually required to obtain detailed results is more quickly and more accurately accomplished by computer than by hand. Since our emphasis is on computation, we have dropped mathematical rigor in favor of intelligibility of the computational methods. However, we have provided the reader with references to standard mathematical textbooks or research papers.

The book assumes a knowledge of mathematics through a first year graduate course in applied mathematics. We have chosen the computer algebra system MACSYMA because it is popular and easy to learn, and some familiarity with MACSYMA is desirable [35]. For the reader who has no experience with MACSYMA we have provided a short introduction in the Appendix.

This book is perhaps best read in front of a computer terminal running MACSYMA. The reader could then enter the programs in this book as BATCH files, and run them on the sample problems. By examining the value of intermediate variables, greater understanding can be gained as to how the methods and programs work. Moreover, we hope that these programs will be useful utilities to research workers in applied mathematics. A note of caution has to be added: As the computational complexity of a problem is increased, e.g. by increasing the number of parameters or the number of equations or the order of truncation, there will come a point where the programs in this book will cease to work, either because of running out of memory or taking too long to run. We suggest that in such cases the reader may extend the usefulness of the program by tailoring it to fit the particular problem at hand.

We have tested the programs in this book on the following versions of MACSYMA: Eunice MACSYMA 308.2 on a VAX 8500 and MACSYMA 310.35 on a SYMBOLICS 3670. The timings which are given at the end of each run are machine dependent and approximate. Even on the same machine, the time for a given run will vary considerably due to "garbage collections" and other aspects of the LISP environment which are invisible to the user. While we have tried to design the programs to run efficiently, the inventive reader can probably improve upon our schemes and is encouraged to do so.

We offer to send an electronic file containing the programs to those readers who have access to BITNET. Our BITNET addresses are currently: RHRY@CRNLVAX5 (for RHR) and URBY@CRNLVAX5 (for DA).

We wish to thank the following people for the help they gave us: Fritz Busse, Tapesh Chakraborty, P.Y. Chen, Vincent Coppola, Richard Cushman, Gerhard Dangelmayr, Larry Fresinsky, Ruediger Gebauer, Jim Geer, Mohammed Golnaraghi, John Guckenheimer, Werner Guettinger, Peter Haug, Tim Healey, Phil Holmes, Herbert Hui, Bill Keith, Jon Len, Alexander Mielke, Frank Moon, Martin Neveling, Ken Ridley, Jan Sanders.

This work was partially supported by the Department of Theoretical and Applied Mechanics of Cornell University, the National Science Foundation, the Mathematical Sciences Institute at Cornell, the Deutsche Forschungsgemeinschaft and the Stiftung Volkswagenwerk.

Richard Rand
Department of Theoretical
and Applied Mechanics
Kimball Hall
Cornell University
Ithaca, NY 14853
USA

Dieter Armbruster
Institut fuer
Informationsverarbeitung
Universitaet Tuebingen
Koestlinstrasse 6
7400 Tuebingen 1
West Germany

May 1987
Ithaca

Introduction

Lindstedt's perturbation method is a classical scheme for obtaining approximate solutions to differential equations which contain a small parameter ϵ. The idea is to expand the solution in a power series in ϵ,

(1)
$$x(t) = x_0(t) + x_1(t) \epsilon + x_2(t) \epsilon^2 + \cdots$$

and to solve for the unknown functions $x_i(t)$ recursively, i.e., in the order $x_0(t)$, $x_1(t)$, $x_2(t)$, ...

Unfortunately, it is generally insufficient to simply substitute eq.(1) into the differential equation to be solved. The shortcoming of such an approach may be illustrated by trying it on an example. We will take van der Pol's equation as our example:

(2)
$$\frac{d^2x}{dt^2} + x + \epsilon (x^2 - 1) \frac{dx}{dt} = 0$$

For any ϵ, this differential equation possesses a periodic solution called a _limit cycle_ (unique to within an arbitrary phase shift in time ([8],§8.7)). It's importance lies in the property that any initial condition (besides the rest state $x = x' = 0$) will eventually lead to the limit cycle, i.e., the limit cycle is globally attracting (for $\epsilon > 0$). Moreover, when $\epsilon = 0$, van der Pol's equation reduces to the simple harmonic oscillator and is easy to solve. Thus we may hope to "perturb off of" the $\epsilon = 0$ solution by substituting the series (1) into equation (2). Keeping terms of $O(\epsilon^2)$, we obtain:

(3) $x_0'' + x_0 + \epsilon [x_1'' + x_1 + x_0'(x_0^2 - 1)]$

$$+ \epsilon^2 [x_2'' + x_2 + x_1'(x_0^2 - 1) + 2 x_0 x_0' x_1] + O(\epsilon^3) = 0$$

Since we want a solution which is valid for all (small) values of ϵ, we equate to zero each of the coefficients of ϵ^n, for $n = 0,1,2,\ldots$. For $n = 0,1$ and 2 we obtain:

(4) $x_0'' + x_0 = 0$

(5) $x_1'' + x_1 = x_0'(1 - x_0^2)$

(6) $x_2'' + x_2 = x_1'(1 - x_0^2) - 2 x_0 x_0' x_1$

Since van der Pol's equation is autonomous (i.e., has no explicit time dependent terms), we can choose the instant $t = 0$ to correspond to any point on the limit cycle. Thus we can choose the initial condition $x'(0) = 0$ without loss of generality. From the expansion (1) we then obtain the initial conditions:

(7) $x_0'(0) = x_1'(0) = x_2'(0) = \cdots = 0$

Eqs.(4) and (7) give

(8) $x_0(t) = B_0 \cos t$

where B_0 is as yet undetermined. Substituting (8) into (5) and using some trig identities, we obtain

(9) $x_1'' + x_1 = \left[\dfrac{B_0^{\,3}}{4} - B_0 \right] \sin t + \dfrac{B_0^{\,3}}{4} \sin 3t$

which has the general solution:

(10) $x_1(t) = \left[\dfrac{B_0^{\,3}}{4} - B_0 \right] (-\dfrac{t}{2} \cos t) - \dfrac{B_0^{\,3}}{32} \sin 3t + B_1 \cos t + A_1 \sin t$

where A_1 and B_1 are arbitrary constants of integration.

Note the presence of the t cos t term in eq.(10). This term grows unbounded as $t \to \infty$. In the final expression for x, eq.(1), this term would appear multiplied by ϵ, i.e., $\epsilon t \cos t$. Thus when t grows to be $O(1/\epsilon)$, this ϵt term becomes $O(1)$ and invalidates the assumption that each term of the series is asymptotically smaller than the preceding term. Inclusion of such unbounded secular or resonance terms prevents the obtained solution from being uniformly valid over the time interval $[0,\infty)$. More simplistically, since we are looking for a periodic solution, we require all (nonperiodic) secular terms to be removed. In the case of eq.(10), this requires that we choose

(11) $B_0 = 2$

So far so good. The procedure fails, however, when we consider eq.(6). First we apply the initial condition (7) to eq.(10), giving:

(12) $A_1 = 3/4$, $x_1(t) = -\frac{1}{4} \sin 3t + \frac{3}{4} \sin t + B_1 \cos t$

Substituting (8) and (12) into (6) and using some trig identities, we obtain

(13) $x_2'' + x_2 = \frac{1}{4} \cos t + 2B_1 \sin t - \frac{3}{2} \cos 3t + 3B_1 \sin 3t + \frac{5}{4} \cos 5t$

Note that the cos t term leads to secular behavior, and that no choice of the constant B_1 will eliminate the problem. Thus the foregoing method, called regular perturbations, fails to yield a solution of the desired form.

Lindstedt's method involves a small modification of regular perturbations, a modification which permits the removal of all secular terms. The idea of the modification can be explained physically by reference to a key distinction between linear and nonlinear dynamical systems. Nonlinear behavior is characterized by a dependence of frequency on amplitude. (Think of the plane pendulum, which obviously takes longer to complete a large amplitude oscillation than a small amplitude oscillation.) In the case of linear oscillations, however, the eigenfrequency is accompanied by an arbitrary amplitude, exemplified by the fact that eigenvectors are never unique but may be multiplied by an arbitrary constant.

Based on this physical reasoning, Lindstedt's method permits the response of the nonlinear ($\epsilon \neq 0$) system to occur at a different frequency from that of the linear ($\epsilon = 0$) system. The idea is to stretch time with the transformation:

(14) $\tau = \omega t$,

where ω, the frequency of the response, is expanded in a power series in ϵ:

(15) $\omega = 1 + k_1 \epsilon + k_2 \epsilon^2 + \cdots$

where the constants k_i are to be found. From a strictly algebraic point of view, (14) represents a change of independent variable which inserts the undetermined constants k_i into the solution, and thereby permits the troublesome secular terms to be removed.

As an example, again consider van der Pol's eq.(2). After the stretch (14), (2) becomes:

$$(16) \qquad \omega^2 \frac{d^2x}{d\tau^2} + x + \epsilon \, (x^2 - 1) \, \omega \, \frac{dx}{d\tau} = 0$$

and after substituting (1) and (15) and collecting terms, we obtain the following equations (cf. eqs.(4)-(6)):

$$(17) \qquad x_0{}'' + x_0 = 0$$

$$(18) \qquad x_1{}'' + x_1 = x_0{}'(1 - x_0{}^2) - 2k_1 \, x_0{}''$$

$$(19) \qquad x_2{}'' + x_2 = x_1{}'(1 - x_0{}^2) - 2 \, x_0 \, x_0{}' \, x_1 - 2k_1 \, x_1{}''$$

$$- (2k_2 + k_1{}^2) \, x_0{}'' + k_1(1 - x_0{}^2) \, x_0{}'$$

where primes now represent derivatives with respect to τ.

As in the case of regular perturbations, eq.(17) has the solution (8). For no secular terms in (18) we find that $B_0 = 2$ and $k_1 = 0$, so that $x_1(\tau)$ is given by (12). When these results are substituted into (19), we obtain:

$$(20) \qquad x_2{}'' + x_2 = (4k_2 + \tfrac{1}{4}) \cos \tau + 2B_1 \sin \tau - \tfrac{3}{2} \cos 3\tau + 3B_1 \sin 3\tau$$

$$+ \tfrac{5}{4} \cos 5\tau$$

When eq.(20) is compared to its regular perturbation counterpart, eq.(13), we see that Lindstedt's method has permitted the removal of secular terms by the choice $B_1 = 0$ and $k_2 = -1/16$.

Computer Algebra

The application of Lindstedt's method to finding limit cycles in the class of equations

$$(21) \qquad\qquad x'' + x + \epsilon\, f(x, x') = 0$$

is conveniently accomplished by the use of computer algebra [35]. We present a MACSYMA program which asks the user to enter the given function $f(x, x')$ and the desired truncation order. Here is a sample run, followed by the program listing. LC is the name of the MACSYMA function which contains the program:


```
LC();

THE D.E. IS OF THE FORM:  X'' + X + E * F(X,X') = 0
ENTER F(X,Y), REPRESENTING X' AS Y
Y*(X^2-1);
                              2
The d.e. is: x'' + x + e ( (x  - 1) y ) = 0
ENTER TRUNCATION ORDER
4;
CHOICES FOR LIMIT CYCLE AMPLITUDE:
1 )   - 2
2 )    2
3 )    0
ENTER CHOICE NUMBER
2;
Done with step 1 of 4
Done with step 2 of 4
Done with step 3 of 4
Done with step 4 of 4
```

$$x = 2 \cos(z) - \frac{(\sin(3\,z) - 3\sin(z))\,e}{4}$$

$$- \frac{(5\cos(5\,z) - 18\cos(3\,z) + 12\cos(z))\,e^2}{96}$$

$$+ \frac{(28\sin(7\,z) - 140\sin(5\,z) + 189\sin(3\,z) - 63\sin(z))\,e^3}{2304} + \ldots$$

$$w = \frac{17\,e^4}{3072} - \frac{e^2}{16} + 1$$

[VAX 8500 TIME = 30 SEC.]

Here is the MACSYMA program listing:

```
/* THIS PROGRAM APPLIES LINDSTEDT'S METHOD TO THE EQUATION:
            X'' + X + E F(X,X') = 0,
ASSUMING A LIMIT CYCLE EXISTS.  CALL IT WITH:  LC();        */
LC():=(
/* input the differential equation */
kill(x,xlist,paramlist),
print("THE D.E. IS OF THE FORM:  X'' + X + E * F(X,X') = 0"),
f:read("ENTER F(X,Y), REPRESENTING X' AS Y"),
print("The d.e. is: x'' + x + e (",f,") = 0"),
f:subst('DIFF(x,z,1)*w,y,f),
/* set up the series expansions */
n:read("ENTER TRUNCATION ORDER"),
w:1,
for i thru n do w:w+k[i]*e^i,
x:b[0]*cos(z),
xlist:[xx[0] = x],
```

```
for i thru n do x:x+xx[i]*e^i,

/* plug into the d.e. and collect terms */

depends(xx,z),

temp1:diff(x,z,2)+x/w^2+e*ev(f,diff)/w^2,

temp1:taylor(temp1,e,0,n),

for i thru n do eq[i]:coeff(temp1,e,i),

/* set up pattern matching rules for use in solving d.e. */

matchdeclare(n1,true),

defrule(c,cos(n1*z),cos(n1*z)/(n1*n1-1)),

defrule(s,sin(n1*z),sin(n1*z)/(n1*n1-1)),

/* load poisson series package and set parameter */

outofpois(dummy),

poislim:100,


/* main loop */

for i:1 thru n do block(

/* trigonometric simplification */

/* efficient alternative to EXPAND(TRIGREDUCE(EXPAND( ))) */

temp1:outofpois(ev(eq[i],xlist,paramlist,diff)),

/* eliminate secular terms */

if i = 1

    then (paramlist:solve(coeff(temp1,sin(z)),b[0]),

        print("CHOICES FOR LIMIT CYCLE AMPLITUDE:"),

        for j:1 thru length(paramlist) do

            print(j,") ",part(paramlist,j,2)),

        r1:read("ENTER CHOICE NUMBER"),

        paramlist:append(solve(coeff(temp1,cos(z)),k[1]),

                [part(paramlist,r1)]))
```

```
        else  paramlist:append(paramlist,

                            solve([coeff(temp1,cos(z)),coeff(temp1,sin(z))],

                            [k[i],b[i-1]]))),

    temp1:expand(ev(temp1,paramlist)),

    xlist:expand(ev(xlist,paramlist)),

    /* output progress */

    print("Done with step",i,"of",n),

    /* exit here if last iteration */

    if i=n then go(end),

    /* solve the d.e. */

    /* efficient alternative to ODE2() */

    temp1:factor(ev(temp1,xx[i] = 0)),

    temp1:applyb1(temp1,c,s),

    temp1:xx[i] = temp1+a[i]*sin(z)+b[i]*cos(z),

    /* fit the initial condition */

    temp2:rhs(temp1),

    temp2:diff(temp2,z),

    temp2:solve(ev(temp2,z:0),a[i]),

    xlist:append(xlist,[ev(temp1,temp2)]),

    /* end of main loop */

    end),


    /* output results */

    w:ev(w,paramlist),

    x:taylor(ev(x,xlist,paramlist),e,0,n-1),

    print("x =",x),

    print("w =",w))$
```

In order to check this program, Figs.1 and 2 compare the order 4 solution generated in the sample run with numerical results obtained by a Runge Kutta scheme. In Fig.1 there is reasonable agreement even though the value of ϵ is as large as 1. The solution has already become unacceptable by $\epsilon = 2$, however (see Fig.2.)

Quadratic Nonlinearities

The foregoing program LC works well for functions $f(x,x')$ (cf. eq.(21)) which, as in the case of van der Pol's equation, do not contain quadratic terms. However, it fails on an example such as

(22) $$x'' + x - \epsilon x'(1 - x + x') = 0$$

for the following reason. The program LC removes secular terms from the x_1 equation and asks the user for a choice of the limit cycle amplitude B_0 on the first pass through the perturbation method. The problem here is that the quadratic terms produce no secular terms on the first pass through the perturbation method. This is due in essence to the identity

$$\cos^2 t = \frac{1}{2} + \frac{1}{2} \cos 2t$$

which contains no secular $\sin t$ or $\cos t$ terms. In order to handle such terms, the program must wait until the x_2 equation to find B_0, i.e. until the second pass through the perturbation method.

This modification can be facilitated by scaling the terms in the differential equation [36]. Let us take the d.e. in the general form:

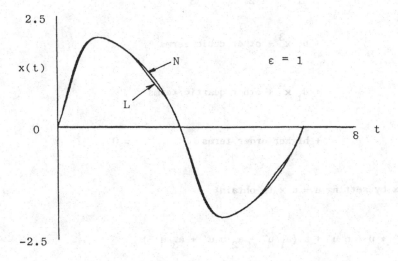

Fig.1. Numerical (=N) and Lindstedt O(4) (=L) solutions
for the limit cycle in van der Pol's eq.(2) for ε = 1.

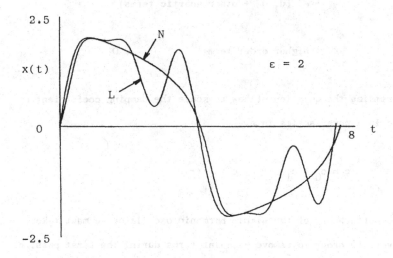

Fig.2. Numerical (= N) and Lindstedt O(4) (= L) solutions
for the limit cycle in van der Pol's eq.(2) for ε = 2.

$$x'' + x + c\,x' + a_1\,x^2 + a_2\,x\,x' + a_3\,x'^2$$

$$+ b_1\,x^3 + \text{other cubic terms}$$

$$+ d_1\,x^4 + \text{other quartic terms}$$

$$+ \text{higher order terms} \qquad\qquad = 0$$

Now we scale x by setting $u = \epsilon\,x$ to obtain:

$$u'' + u + c\,u' + \epsilon\,(a_1\,u^2 + a_2\,u\,u' + a_3\,u'^2)$$

$$+ \epsilon^2\,(b_1\,u^3 + \text{other cubic terms})$$

$$+ \epsilon^3\,(d_1\,u^4 + \text{other quartic terms})$$

$$+ \text{higher order terms} \qquad\qquad = 0$$

There remains the question of how to scale the damping coefficient c. Let us expand c in a power series in ϵ:

$$c = c_0 + c_1\,\epsilon + c_2\,\epsilon^2 + \cdots$$

Now in order to perturb off of the simple harmonic oscillator we must take $c_0 = 0$. Moreover, in order to remove resonant terms during the first pass, c_1 must be chosen to be zero. (If c_1 were not chosen to be zero, the perturbation method would fail to obtain a limit cycle.) Thus we scale the linear damping term to be of order ϵ^2.

In summary, we require the terms to scale as follows:

$$\text{linear: } O(\epsilon^2)$$

$$\text{quadratic: } O(\epsilon),$$

$$\text{cubic: } O(\epsilon^2),$$

$$\text{quartic: } O(\epsilon^3), \text{ etc.}$$

For example in eq. (22), set $\tilde{\epsilon} = \epsilon^{1/2}$ and $\tilde{x} = \tilde{\epsilon} x$ to obtain

(23) $$\tilde{x}'' + \tilde{x} - \tilde{\epsilon}^2 \tilde{x}' + \tilde{\epsilon} \tilde{x} \tilde{x}' - \tilde{\epsilon} \tilde{x}'^2 = 0$$

The following program LC2 is a generalization of LC for the system:

(24.1) $$x' = -y + \epsilon f(x,y)$$

(24.2) $$y' = x + \epsilon g(x,y)$$

where $f(x,y)$ and $g(x,y)$ are assumed to be in the form:

$$\text{quadratic} + \epsilon [\text{linear} + \text{cubic}] + \epsilon^2 \text{ quartic} + \cdots$$

Here is a sample run based on eq. (23) in the form (24), followed by the program listing:

```
LC2();
The d.e.'s are of the form x' = -y + e*f, y' = x + e*g
where f,g are of the form:
        quadratic + e*(linear + cubic) + e^2*(quartic) + ...
Enter f:
0;
f = 0
Enter g:
E*Y-Y^2-X*Y;

        2
g = - y  - x y + e y
```

```
Enter truncation order:
4;
Done with step 1 of 4
Choices for limit cycle amplitude:
1 )    0
2 )   - 2
3 )    2
Enter choice number
3;
Done with step 2 of 4
Done with step 3 of 4
Done with step 4 of 4
```

$$x = 2\cos(z) + \frac{(4\sin(z) + 2\cos(2z) - 2\sin(2z) + 6)\,e}{3}$$

$$+ (62\cos(z) - \sin(z) + 32\cos(2z) + 32\sin(2z) + 3\cos(3z) - 21\sin(3z))$$

$$e^2/36 + \ldots$$

$$y = 2\sin(z) - \frac{(4\cos(z) - 4\cos(2z) - 4\sin(2z))\,e}{3}$$

$$+ (\cos(z) + 2\sin(z) - 64\cos(2z) + 64\sin(2z) + 63\cos(3z) + 9\sin(3z))$$

$$e^2/36 + \ldots$$

$$w = -\frac{113\,e^4}{108} - \frac{5\,e^2}{6} + 1$$

```
[VAX 8500 TIME = 100 SEC.]
```

Here is the MACSYMA program listing:

```
LC2():=(

/* input the differential equation */

kill(x,y,xylist,paramlist),

print("The d.e.'s are of the form x' = -y + e*f, y' = x + e*g"),

print("where f,g are of the form:

         quadratic + e*(linear + cubic) + e^2*(quartic) + ..."),

f:read("Enter f:"),

print("f =",f),
```

```
g:read("Enter g:"),

print("g =",g),

/* set up the series expansions */

n:read("Enter truncation order:"),

k[0]:1,

k[1]:0,

w:sum(k[i]*e^i,i,0,n),

xy:[x:b[0]*cos(z)+sum(xx[i](z)*e^i,i,1,n),

   y:b[0]*sin(z)+sum(yy[i](z)*e^i,i,1,n)],

xylist:[xx[0](z)=b[0]*cos(z),

        yy[0](z)=b[0]*sin(z)],

/* plug into d.e.'s and collect terms */

temp1:[-diff(x,z)*w-y+e*ev(f,diff),-diff(y,z)*w+x+e*ev(g,diff)],

temp2:taylor(temp1,e,0,n),

for i:1 thru n do eq[i]:coeff(temp2,e,i),


/* main loop */

for i:1 thru n do block(

/* trigonometric simplification */

temp3:expand(trigreduce(expand(ev(eq[i],xylist,paramlist,diff)))),

/* eliminate secular terms */

if i=1 then (temp4:temp3, go(skip_to_here_first_time))

      else newparamlist:

             solve([coeff(part(temp3,1),sin(z))-coeff(part(temp3,2),cos(z)),

                    coeff(part(temp3,1),cos(z))+coeff(part(temp3,2),sin(z))],

                 [b[i-2],k[i]]),

if i=2 then (paramlist:newparamlist,

             print("Choices for limit cycle amplitude:")),
```

```
        for j:1 thru length(paramlist) do

            print(j,")  ",part(paramlist,j,1,2)),

        r1:read("Enter choice number"),

        paramlist:part(paramlist,r1))

    else paramlist:append(paramlist,newparamlist),

temp4:expand(ev(temp3,paramlist)),

xylist:expand(ev(xylist,paramlist)),

skip_to_here_first_time,

/* output progress */

print("Done with step",i,"of",n),

/* exit here if last iteration */

if i=n then go(end),

/* solve the d.e.'s */

temp4a:subst(dummy(z),yy[i](z),temp4),

atvalue(dummy(z),z=0,0),

temp5:desolve(temp4a,[xx[i](z),dummy(z)]),

temp5a:subst(yy[i](z),dummy(z),temp5),

temp5b:subst(b[i],xx[i](0),temp5a),

xylist:append(xylist,[temp5b]),

/* end of main loop */

end),


/* output results */

w:ev(w,paramlist),

soln:taylor(ev([x,y],xylist,paramlist),e,0,n-2),

print("x =",part(soln,1)),

print("y =",part(soln,2)),

print("w =",w))$
```

As a check on the program and the method, we compare the results of LC2 with numerical integration on eq.(22). Fig.3 shows the limit cycle of eq.(22) for $\epsilon = 0.16$ (i.e. $\tilde{\epsilon} = 0.4$ in the perturbation scheme, see eq.(23.)) The curves in Fig.3 were obtained from the previous run (truncation of order 4), as well as from runs with truncation orders 6 and 8. In addition, Fig.3 displays numerical results generated by a Runge Kutta scheme. Note how, for this choice of ϵ, greater accuracy is achieved by taking more terms in the series.

Hopf Bifurcation

As an application of the program LC2, we investigate the Hopf bifurcation [36]. Whenever an equilibrium point of the focus type (i.e. having complex eigenvalues) changes its stability as a result of parameter changes in the differential equations, a limit cycle is generically born. In order to use Lindstedt's method to see how this occurs, we consider the most general version of eqs.(24) up to cubic terms, i.e. we expand f and g in Taylor series. We also include damping terms with coefficient μ so that varying μ through zero will change the stability of the equilibrium, producing the desired result:

$$(25.1) \quad x' = -y + \epsilon^2 \mu x + \epsilon \left(\frac{fxx}{2} x^2 + fxy \, x \, y + \frac{fyy}{2} y^2 \right)$$

$$+ \epsilon^2 \left(\frac{fxxx}{6} x^3 + \frac{fxxy}{2} x^2 y + \frac{fxyy}{2} x \, y^2 + \frac{fyyy}{6} y^3 \right) + \cdots$$

$$(25.2) \quad y' = x + \epsilon^2 \mu y + \epsilon \left(\frac{gxx}{2} x^2 + gxy \, x \, y + \frac{gyy}{2} y^2 \right)$$

$$+ \epsilon^2 \left(\frac{gxxx}{6} x^3 + \frac{gxxy}{2} x^2 y + \frac{gxyy}{2} x \, y^2 + \frac{gyyy}{6} y^3 \right) + \cdots$$

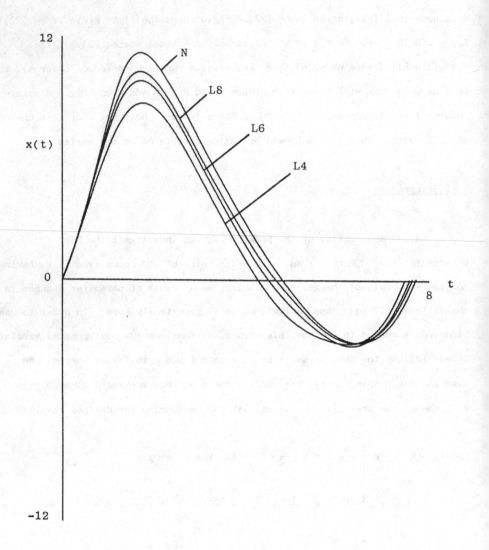

Fig.3. Numerical (= N) and Lindstedt O(8), O(6) and O(4)
(= L8, L6 and L4 respectively) solutions for the limit
cycle of eq.(22) for ε = 0.16 (i.e. ε̃ = 0.4).

Here is the run:

LC2();

The d.e.'s are of the form x' = -y + e*f, y' = x + e*g
where f,g are of the form:
          quadratic + e*(linear + cubic) + e^2*(quartic) + ...
Enter f:
e*(fyyy*y^3/6+fxyy*x*y^2/2+fxxy*x^2*y/2+fxxx*x^3/6)+fyy*y^2/2+fxy*x*y
+fxx*x^2/2+e*mu*x;

$$f = e\left(\frac{fyyy\ y^3}{6} + \frac{fxyy\ x\ y^2}{2} + \frac{fxxy\ x^2\ y}{2} + \frac{fxxx\ x^3}{6}\right) + \frac{fyy\ y^2}{2} + fxy\ x\ y + \frac{fxx\ x^2}{2}$$

$$+ e\ mu\ x$$

Enter g:
e*(gyyy*y^3/6+gxyy*x*y^2/2+gxxy*x^2*y/2+gxxx*x^3/6)+gyy*y^2/2+gxy*x*y
+gxx*x^2/2+e*mu*y;

$$g = e\left(\frac{gyyy\ y^3}{6} + \frac{gxyy\ x^2\ y}{2} + \frac{gxxy\ x^2\ y}{2} + \frac{gxxx\ x^3}{6}\right) + \frac{gyy\ y^2}{2} + gxy\ x\ y + e\ mu\ y$$

$$+ \frac{gxx\ x^2}{2}$$

Enter truncation order:
2;
Done with step 1 of 2
Choices for limit cycle amplitude:
1)   0
2)   - 4 sqrt(- mu/(gyyy - gxy gyy + fyy gyy - gxx gxy + gxxy - fxx gxx

               + fxy fyy + fxyy + fxx fxy + fxxx))

3)   4 sqrt(- mu/(gyyy - gxy gyy + fyy gyy - gxx gxy + gxxy - fxx gxx

               + fxy fyy + fxyy + fxx fxy + fxxx))

Enter choice number
3;
Done with step 2 of 2

x = 4 cos(z) sqrt(- mu/(gyyy + (- gxy + fyy) gyy - gxx gxy + gxxy - fxx gxx

               + fxy fyy + fxyy + fxx fxy + fxxx)) + . . .

y = 4 sin(z) sqrt(- mu/(gyyy + (- gxy + fyy) gyy - gxx gxy + gxxy - fxx gxx

               + fxy fyy + fxyy + fxx fxy + fxxx)) + . . .

$$w = e^2 \, (2\, gyy^2 + (5\, gxx - 5\, fxy)\, gyy - 3\, gxyy + 2\, gxy^2 + (-\,fyy - 5\, fxx)\, gxy$$

$$-\, 3\, gxxx + 5\, gxx^2 - fxy\, gxx + 3\, fyyy + 5\, fyy^2 + 5\, fxx\, fyy + 2\, fxy^2 + 3\, fxxy$$

$$+\, 2\, fxx^2)\, mu/(3\, gyyy + (3\, fyy - 3\, gxy)\, gyy - 3\, gxx\, gxy + 3\, gxxy - 3\, fxx\, gxx$$

$$+\, 3\, fxy\, fyy + 3\, fxyy + 3\, fxx\, fxy + 3\, fxxx) + 1$$

[VAX 8500 TIME = 331 SEC.]

The significance of this computation lies in the requirement that the limit cycle amplitude be real. The limit cycle amplitude is given by the expression:

(26)   limit cycle amplitude $= 4\,(-\mu/S)^{1/2}$.

where

(27) $S = gyyy + gxxy + fxxx + fxyy + fyy\, gyy + fxy\,(fxx + fyy)$

$$-\, fxx\, gxx - gxy\,(gxx + gyy)$$

In particular this requires that the sign of S be opposite to the sign of μ. If $S = 0$ the computation is indecisive and offers no information. See Fig.4 for the two cases of subcritical and supercritical Hopf bifurcations, in which the stability of the limit cycle is respectively unstable and stable.

Note that the form of the linear terms in eqs.(25) involves the symmetric appearance of μ in both the x' and y' equations. A general system of the form

(28) $x' = a\, x + b\, y + \cdots, \quad y' = c\, x + d\, y + \cdots$

can be transformed into

$$u' = -\,\Omega\, v + \mu\, u + \cdots, \quad v' = \Omega\, u + \mu\, v + \cdots$$

SUPERCRITICAL CASE, S < 0

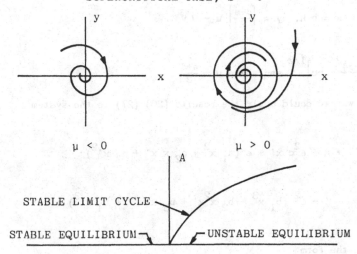

SUBCRITICAL CASE, S > 0

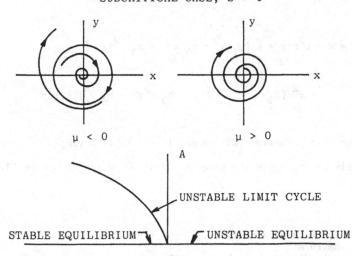

Fig.4. Supercritical and subcritical Hopf Bifurcations.
A = amplitude of limit cycle.

via the linear transformation [36]

(29) $x = b\ u,\quad y = \left[\dfrac{d-a}{2}\right] u - \Omega\ v$

where $\Omega = \left[-\dfrac{(d-a)^2}{4} - bc\right]^{1/2}$ and where $\mu = \dfrac{a+d}{2}$.

In this way we could apply the results (26),(27) to the system

(30) $x'' + x + \epsilon^2 c\ x' + \epsilon\ (a_1\ x^2 + a_2\ x\ x' + a_3\ x'^2)$

$\qquad\qquad + \epsilon^2\ (b_1\ x^3 + b_2\ x^2 x' + b_3\ x\ x'^2 + b_4\ x'^3) + \cdots$

by writing it in the form:

(31.1) $x' = -\ y$

(31.2) $y' = x - \epsilon^2 c\ y + \epsilon\ (a_1\ x^2 - a_2\ x\ y + a_3\ y^2)$

$\qquad\qquad + \epsilon^2\ (b_1\ x^3 - b_2\ x^2\ y + b_3\ x\ y^2 - b_4\ y^3) + \cdots$

and then using the transformation (29) to get it in the form (25).

Alternately we can apply the program LC2 directly to the system (31):

LC2();

The d.e.'s are of the form x' = -y + e*f, y' = x + e*g
where f,g are of the form:
        quadratic + e*(linear + cubic) + e^2*(quartic) + ...
Enter f:
0;
f = 0
Enter g:
-c*y*e+(a1*x^2-a2*x*y+a3*y^2)+e*(b1*x^3-b2*x^2*y+b3*x*y^2-b4*y^3);

$$g = e\ (-\ b4\ y^3 + b3\ x\ y^2 - b2\ x^2\ y + b1\ x^3) + a3\ y^2 - a2\ x\ y - c\ e\ y + a1\ x^2$$

```
Enter truncation order:
2;
Done with step 1 of 2
Choices for limit cycle amplitude:
1 )   0

                            c
2 )    - 2 sqrt(- ---------------------)
                  3 b4 + b2 - a2 a3 - a1 a2


                          c
3 )    2 sqrt(- ---------------------)
                3 b4 + b2 - a2 a3 - a1 a2

Enter choice number
3;
Done with step 2 of 2
                          c
x = 2 cos(z) sqrt(- ---------------------) + . . .
                    3 b4 + b2 - a2 a3 - a1 a2


                          c
y = 2 sin(z) sqrt(- ---------------------) + . . .
                    3 b4 + b2 - a2 a3 - a1 a2


            2              2        2       2
      (3 b3 + 9 b1 - 4 a3  - 10 a1 a3 - a2  - 10 a1 ) c e
w = 1 - ---------------------------------------------------
            18 b4 + 6 b2 - 6 a2 a3 - 6 a1 a2

[VAX 8500 TIME = 74 SEC.]
```

From this run we can conclude that for small ϵ eq. (30) possesses a limit cycle if the quantity

$$(32) \qquad\qquad 3\, b_4 + b_2 - a_2\, a_3 - a_1\, a_2$$

has the opposite sign to that of the damping c. For example, in the case of van der Pol's equation (2), where $c = -1$, $b_2 = 1$ and the other coefficients are zero, the quantity (32) becomes unity and the limit cycle amplitude is correctly evaluated as 2. Note also that the ϵ^2 contribution to the frequency ω vanishes for van der Pol's equation. (Cf. eq. (15) where $k_1 = 0$ and where ϵ must be replaced by ϵ^2 for comparison with the foregoing run.)

Exercises

1. Find an equation of the form

$$x'' + x + \epsilon \, f(x,x') = 0$$

which exhibits a limit cycle with frequency $\omega = 1 + k_1 \epsilon + \cdots$, where $k_1 \neq 0$.
Note that the examples used in this Chapter, namely van der Pol's equation (2),
eq.(22), and eq.(30), all have $k_1 = 0$.
Hint: Take $a_1 = a_2 = a_3 = 0$ in eq.(30).

2. We have dealt exclusively with <u>autonomous</u> systems in this Chapter. Write a
program to apply Lindstedt's method to <u>nonautonomous</u> equations of the form

$$x'' + x + \epsilon \, f(x,x',t) = 0.$$

As an example, consider the forced damped Duffing equation,

$$x'' + x + \epsilon \, \delta \, x' + \epsilon \, \alpha \, x^3 = \epsilon \, \gamma \, \cos \omega t$$

where $\omega = 1 + k_1 \epsilon + k_2 \epsilon^2 + \cdots$ is <u>given</u>.
Discussion: First we set $\tau = \omega t$ to get

$$\omega^2 \frac{d^2 x}{d\tau^2} + x + \epsilon \, \delta \, \omega \frac{dx}{d\tau} + \epsilon \, \alpha \, x^3 = \epsilon \, \gamma \, \cos \tau$$

and then we expand $x = x_0(\tau) + x_1(\tau) \epsilon + \cdots$. We look for solutions with
period 2π in τ, i.e. with the same period as the forcing function.

In order to eliminate secular terms in the x_1 equation we can now choose A_0 and B_0. (Note that the k_i's are viewed as being given in this formulation.) Similarly the complementary solution for the x_i equation can be chosen as $x_{i_{comp}} = A_i \sin \tau + B_i \cos \tau$ and an appropriate selection of A_i and B_i will eliminate secular terms in the x_{i+1} equation.

Unfortunately the algebraic equations on A_0 and B_0 (or more generally on A_i and B_i) are usually too complicated to solve in closed form. Transformation to polar coordinates

$$A_0 = R_0 \cos \theta_0, \quad B_0 = R_0 \sin \theta_0$$

usually helps. E.g. in the case of Duffing's equation, A_0 and B_0 satisfy

$$- 2 B_0 k_1 + A_0 \delta + \frac{3}{4} \alpha B_0^3 + \frac{3}{4} \alpha A_0^2 B_0 = \gamma$$

$$- 2 A_0 k_1 - B_0 \delta + \frac{3}{4} \alpha A_0^3 + \frac{3}{4} \alpha A_0 B_0^2 = 0$$

which become in polar coordinates,

$$\frac{3}{4} \alpha R_0^3 - 2 k_1 R_0 = \gamma \sin \theta_0$$

$$\delta R_0 = \gamma \cos \theta_0$$

which can be combined to give

$$\left[\frac{3}{4} \alpha R_0^3 - 2 k_1 R_0\right]^2 + \left[\delta R_0\right]^2 = \gamma^2$$

3. Mathieu's equation,

$$x'' + (\delta + \epsilon \cos t) x = 0$$

is both <u>linear</u> and <u>nonautonomous</u>. The problem here is to find the coefficients δ_i in the expansion

$$\delta = \frac{n^2}{4} + \delta_1 \epsilon + \delta_2 \epsilon^2 + \cdots, \quad n = 0,1,2,\cdots$$

such that the differential equation exhibits periodic solutions. This expansion represents a <u>transition curve</u> in the δ-ϵ parameter plane (hence the phrase <u>parametric excitation</u>) which separates regions of bounded behavior from regions of unbounded behavior. See e.g. [44].

Write a program to accomplish this task. Note that since the equation is linear, no frequency-amplitude relation is expected and no stretch $\tau = \omega t$ is needed, so that <u>regular perturbations</u> will work on this problem.

After expanding $x = x_0 + x_1 \epsilon + \cdots$, separately choose $x_0 = \sin \frac{n}{2} t$ and then $x_0 = \cos \frac{n}{2} t$, as each gives a distinct transition curve. Find the δ_i by elimination of secular terms. See [35].

4. Apply the transformation (29) to eqs.(31) to get them in the form (25). Then show that the results (26) and (27) agree with results of the last run in the Chapter on eqs.(31) directly.

CHAPTER 2

CENTER MANIFOLDS

Introduction

Center manifold theory [6] is a method which uses power series
expansions in the neighborhood of an equilibrium point in order to reduce the
dimension of a system of ordinary differential equations. The method involves
restricting attention to an invariant subspace (the center manifold) which
contains all of the essential behavior of the system in the neighborhood of the
equilibrium point as $t \to \infty$.

This method is applicable to systems which, when linearized about an
equilibrium point, have some eigenvalues which have zero real part, and others
which have negative real part. We assume that no eigenvalues have positive
real part, since in such a case the center manifold will not be attractive as
$t \to \infty$.

Under such assumptions the components of the solution of the linearized
equations which correspond to those eigenvalues with negative real part will
decay as $t \to \infty$, and hence the motion of the linearized system will
asymptotically approach the space S_1 spanned by the eigenvectors corresponding
to those eigenvalues with zero real part. The center manifold theorem [6]
assures us that this picture (which is so far based on the linearized
equations) extends to the full nonlinear equations, as follows:

There exists a (generally curved) subspace S_2 (the center manifold) which is tangent to the (flat) subspace S_1 at the equilibrium point, and which is invariant under the flow generated by the nonlinear equations. All solutions which start sufficiently close to the equilibrium point will tend asymptotically to the center manifold. In particular, the theorem states that the stability of the equilibrium point in the full nonlinear equations is the same as its stability when restricted to the flow on the center manifold ([6], p.4). Moreover, any additional equilibrium points or limit cycles which occur in a neighborhood of the given equilibrium point on the center manifold are guaranteed to exist in the full nonlinear equations ([6], p.29).

Thus center manifold theory allows us to eliminate algebraic complications which are due to unessential behavior and to focus on the motion in the center manifold which contains the critical information concerning the system's local stability and bifurcation of steady state behavior. In doing so we omit reference to how the system gets onto the center manifold, but rather we are assured of the asymptotic stability of the center manifold. Since the eigenvalues associated with the linearized flow in the center manifold have zero real part, the study of the motion in the center manifold must still be accomplished and may not be an easy task. In this Chapter we will be concerned with the question of how to reduce a system to its center manifold and we will leave for later Chapters treatment of the flow on the center manifold.

The method will be illustrated by reference to the following example [31] of a system of three differential equations:

(1) $x' = y$

(2) $y' = -x - x z$

(3) $z' = -z + \alpha x^2$

Here we have a feedback control system in which the spring constant in the simple harmonic oscillator of eqs.(1),(2) is given by $1 + z$. The control variable z is governed by eq.(3) and involves the feedback term αx^2. We wish to know the stability of the equilibrium at the origin.

Although the system (1)-(3) is 3-dimensional, we can use center manifold theory to reduce the stability question to the study of a 2-dimensional system. Linearizing about the origin, we see that the x-y plane is associated with a pair of pure imaginary eigenvalues, while the z axis corresponds to an eigenvalue of -1. Thus the center manifold is a 2-dimensional surface which is tangent to the x-y plane at the origin.

In order to obtain an approximate expression for the center manifold, we write z as a function of x and y, and expand in a power series. Since the center manifold is tangent to the x-y plane, we begin the power series with quadratic terms:

(4) $$z = a_{2,0}\, x^2 + a_{1,1}\, x y + a_{0,2}\, y^2 + \cdots$$

The requirement that the center manifold be invariant is satisfied by substituting (4) into (3):

(5) $$(2\, a_{2,0}\, x + a_{1,1}\, y)\, x' + (2\, a_{0,2}\, y + a_{1,1}\, x)\, y' + \cdots =$$

$$- (a_{2,0}\, x^2 + a_{1,1}\, x y + a_{0,2}\, y^2) + \alpha x^2 + \cdots$$

Note that (5) involves the derivatives x' and y'. Substitution of expressions for these from (1) and (2) gives:

(6) $(2 a_{2,0} x + a_{1,1} y) y + (2 a_{0,2} y + a_{1,1} x)(- x - x z) + \cdots =$

$$- (a_{2,0} x^2 + a_{1,1} x y + a_{0,2} y^2) + \alpha x^2 + \cdots$$

This last step once again introduced z into the computation (since y' depended on z). So we again eliminate z by substituting (4) into (6). Collecting terms and neglecting cubic and higher order terms, this gives:

(7) $(a_{2,0} - a_{1,1} - \alpha) x^2 + (2 a_{2,0} - 2 a_{0,2} + a_{1,1}) x y$

$$+ (a_{1,1} + a_{0,2}) y^2 + \cdots = 0$$

Equating the coefficients of $x^i y^j$ equal to zero, and solving the three resulting equations for the $a_{i,j}$'s, we obtain:

(8) $a_{2,0} = \frac{3}{5} \alpha , \quad a_{1,1} = - \frac{2}{5} \alpha , \quad a_{0,2} = \frac{2}{5} \alpha$

Substitution in (4) gives the following first approximation to the center manifold:

(9) $z = \frac{3}{5} \alpha x^2 - \frac{2}{5} \alpha x y + \frac{2}{5} \alpha y^2 + \cdots$

We may now substitute (9) into eqs. (1),(2) in order to obtain approximate equations for the flow on the center manifold:

(10) $x' = y$

(11) $y' = - x - \frac{3}{5} \alpha x^3 + \frac{2}{5} \alpha x^2 y - \frac{2}{5} \alpha x y^2 + \cdots$

Although approximate, eqs.(10),(11) contain the answer to the question of the stability of the origin in the original system (1)-(3). The point of center manifold theory is that eqs.(10),(11) are easier to analyze than eqs.(1)-(3). We will return to the analysis of eqs.(10),(11) in the next Chapter on normal forms.

Computer Algebra

The calculation of center manifolds involves the manipulation of truncated power series and is readily performed using computer algebra. We present a MACSYMA program to accomplish such computations. The program listing follows a sample run on the preceding example:


```
CM();

ENTER NO. OF EQS.
3;
ENTER DIMENSION OF CENTER MANIFOLD
2;

THE D.E.'S MUST BE ARRANGED SO THAT THE FIRST 2 EQS.
REPRESENT THE CENTER MANIFOLD.  I.E. ALL ASSOCIATED
EIGENVALUES ARE ZERO OR HAVE ZERO REAL PARTS.

ENTER SYMBOL FOR VARIABLE NO. 1
X;
ENTER SYMBOL FOR VARIABLE NO. 2
Y;
ENTER SYMBOL FOR VARIABLE NO. 3
Z;
ENTER ORDER OF TRUNCATION
2;
ENTER RHS OF EQ. 1
D x /DT =
Y;

ENTER RHS OF EQ. 2
D y /DT =
-X-X*Z;

ENTER RHS OF EQ. 3
D z /DT =
-Z+ALPHA*X^2;
```

$$\frac{dx}{dt} = y$$

$$\frac{dy}{dt} = -xz - x$$

$$\frac{dz}{dt} = \text{alpha } x^2 - z$$

CENTER MANIFOLD:

$$[z = \frac{2 \text{ alpha } y^2}{5} - \frac{2 \text{ alpha } x y}{5} + \frac{3 \text{ alpha } x^2}{5}]$$

FLOW ON THE C.M.:

$$[\frac{dx}{dt} = y, \ \frac{dy}{dt} = -x (\frac{2 \text{ alpha } y^2}{5} - \frac{2 \text{ alpha } x y}{5} + \frac{3 \text{ alpha } x^2}{5}) - x]$$

[VAX 8500 TIME = 6 SEC.]

Here is the MACSYMA listing for the program CM:

```
CM():=(
/* INPUT PROBLEM */
N:READ("ENTER NO. OF EQS."),
M:READ("ENTER DIMENSION OF CENTER MANIFOLD"),
PRINT("THE D.E.'S MUST BE ARRANGED SO THAT THE FIRST",M,"EQS."),
PRINT("REPRESENT THE CENTER MANIFOLD.  I.E. ALL ASSOCIATED"),
PRINT("EIGENVALUES ARE ZERO OR HAVE ZERO REAL PARTS."),
FOR I:1 THRU N DO
   X[I]:READ("ENTER SYMBOL FOR VARIABLE NO.",I),
L:READ("ENTER ORDER OF TRUNCATION"),
FOR I:1 THRU N DO (
   PRINT("ENTER RHS OF EQ.",I),
   PRINT("D",X[I],"/DT ="),
```

```
   G[I]:READ()),
/* SET UP D.E.'S */
FOR I:1 THRU N DO
   DEPENDS(X[I],T),
FOR I:1 THRU N DO
   (EQ[I]:DIFF(X[I],T)=G[I],
    PRINT(EQ[I])),
/* FORM POWER SERIES */
SUB:MAKELIST(K[I],I,1,M),
VAR:PRODUCT(X[I]^K[I],I,1,M),
UNK:[],
FOR P:M+1 THRU N DO(
   TEMP:A[P,SUB]*VAR,
   FOR I:1 THRU M DO
      TEMP:SUM(EV(TEMP,K[I]=J),J,O,L),
   TEMP2:TAYLOR(TEMP,MAKELIST(X[I],I,1,M),O,L),
   /* REMOVE CONSTANT AND LINEAR TERMS */
   TEMP3:TEMP2-PART(TEMP2,1)-PART(TEMP2,2),
   SOLN[P]:EXPAND(TEMP3),
   /* PREPARE LIST OF UNKNOWNS */
   SETXTO1:MAKELIST(X[I]=1,I,1,M),
   /* TURN SUM INTO A LIST */
   UNKN[P]:SUBST("[","+",EV(TEMP3,SETXTO1)),
   UNK:APPEND(UNK,UNKN[P])),
SOL:MAKELIST(X[P]=SOLN[P],P,M+1,N),
/* SUBSTITUTE INTO D.E.'S */
CMDE:MAKELIST(EQ[I],I,1,M),
REST:MAKELIST(LHS(EQ[I])-RHS(EQ[I]),I,M+1,N),
TEMP4:EV(REST,SOL,DIFF),
```

```
TEMP5:EV(TEMP4,CMDE,DIFF),

TEMP6:EV(TEMP5,SOL),

TEMP7:TAYLOR(TEMP6,MAKELIST(X[I],I,1,M),0,L),

/* COLLECT TERMS */

COUNTER:1,

/* MAKE LIST OF TERMS */

TERMS:SUBST("[","+",SOLN[N]),

TERMS:EV(TERMS,A[DUMMY,SUB]:=1),

FOR I:1 THRU N-M DO (

    EXP[I]:EXPAND(PART(TEMP7,I)),

    FOR J:1 THRU LENGTH(TERMS) DO(

        COND[COUNTER]:RATCOEF(EXP[I],PART(TERMS,J)),

        COUNTER:COUNTER+1)),

CONDS:MAKELIST(COND[I],I,1,COUNTER-1),

/* SOLVE FOR CENTER MANIFOLD */

ACOEFFS:SOLVE(CONDS,UNK),

CENTERMANIFOLD:EV(SOL,ACOEFFS),

PRINT("CENTER MANIFOLD:"),

PRINT(CENTERMANIFOLD),

/* GET FLOW ON CM */

CMDE2:EV(CMDE,CENTERMANIFOLD),

PRINT("FLOW ON THE C.M.:"),

PRINT(CMDE2))$
```

Systems with Damping

Since the center manifold must contain a linearized system with
eigenvalues equal to zero or having zero real part, it would appear that the
method is inapplicable to the systems which have damping. Carr [6] has shown,
however, that such cases can be handled by embedding the given system in a
larger system which contains the damping as an additional dependent variable.
For example, consider the previous example when damping is added:

(12) $x' = \mu x + y$

(13) $y' = -x + \mu y - x z$

(14) $z' = -z + \alpha x^2$

Now to these equations we add the dummy equation:

(15) $\mu' = 0$

The system (12)-(15) is now a 4-dimensional system with a 3-dimensional
center manifold. The damping terms μx and μy are now nonlinear (quadratic)
terms in the neighborhood of the origin $x = y = z = \mu = 0$. Naturally the
results of the method are only valid for small values of μ.

In order to illustrate the procedure, we apply the program CM to this
example:

CM();

ENTER NO. OF EQS.
4;
ENTER DIMENSION OF CENTER MANIFOLD
3;

THE D.E.'S MUST BE ARRANGED SO THAT THE FIRST 3 EQS.
REPRESENT THE CENTER MANIFOLD.   I.E. ALL ASSOCIATED
EIGENVALUES ARE ZERO OR HAVE ZERO REAL PARTS.

ENTER SYMBOL FOR VARIABLE NO. 1
MU;
ENTER SYMBOL FOR VARIABLE NO. 2
X;
ENTER SYMBOL FOR VARIABLE NO. 3
Y;
ENTER SYMBOL FOR VARIABLE NO. 4
Z;
ENTER ORDER OF TRUNCATION
3;
ENTER RHS OF EQ. 1
D mu /DT =
O;

ENTER RHS OF EQ. 2
D x /DT =
MU*X+Y;

ENTER RHS OF EQ. 3
D y /DT =
MU*Y-X-X*Z;

ENTER RHS OF EQ. 4
D z /DT =
-Z+ALPHA*X^2;

```
dmu
--- = 0
dt

dx
-- = y + mu x
dt

dy
-- = - x z + mu y - x
dt

dz        2
-- = alpha x  - z
dt
```

CENTER MANIFOLD:

$$[z = - \frac{28 \text{ alpha mu } y^2}{25} + \frac{2 \text{ alpha } y^2}{5} + \frac{8 \text{ alpha mu } x \, y}{25} - \frac{2 \text{ alpha } x \, y}{5}$$

$$- \frac{22 \text{ alpha mu } x^2}{25} + \frac{3 \text{ alpha } x^2}{5}]$$

FLOW ON THE C.M.:

$$[\frac{dmu}{dt} = 0, \quad \frac{dx}{dt} = y + mu \, x, \quad \frac{dy}{dt} = - x \, (- \frac{28 \text{ alpha mu } y^2}{25} + \frac{2 \text{ alpha } y^2}{5}$$

$$+ \frac{8 \text{ alpha mu } x \, y}{25} - \frac{2 \text{ alpha } x \, y}{5} - \frac{22 \text{ alpha mu } x^2}{25} + \frac{3 \text{ alpha } x^2}{5}) + mu \, y - x]$$

[VAX 8500 TIME = 51 SEC.]

That is, for small values of μ, the flow near the origin on the center manifold of the system (12)-(15) is given by:

(16) $x' = y + \mu x$

(17) $y' = -x + \mu y + \alpha x \left[-\frac{3}{5} x^2 + \frac{2}{5} x y - \frac{2}{5} y^2 + \mu \left[\frac{22}{25} x^2 - \frac{8}{25} x y + \frac{28}{25} y^2 \right] \right]$

Here the origin is unstable for $\mu > 0$ and stable for $\mu < 0$. Thus there is a Hopf bifurcation at $\mu = 0$. Let us use the Lindstedt method program LC2 of Chapter 1 to approximate the size of the limit cycle as a function of the parameters α and μ. In order to put the problem in a form suitable for treatment by LC2, we scale x, y and μ as follows:

(18) $X = \epsilon x, \quad Y = - \epsilon y, \quad M = \epsilon^2 \mu$

whereupon eqs. (16), (17) become:

(19) $X' = - Y + M \epsilon^2 X$

(20) $Y' = X + M \epsilon^2 Y + \alpha \epsilon^2 X \left[\frac{3}{5} X^2 + \frac{2}{5} X Y + \frac{2}{5} Y^2 \right] + \cdots$

We continue with the application of the program LC2 to eqs.(19),(20):

LC2();

The d.e.'s are of the form x' = -y + e*f, y' = x + e*g

where f,g are of the form:

          quadratic + e*(linear + cubic) + e^2*(quartic) + ...

Enter f:
M*E*X;

f = e m x

Enter g:
M*E*Y+ALPHA*E*X*(3*X^2/5+2*X*Y/5+2*Y^2/5);

$$g = alpha\ e\ x\ (\frac{2\ y^2}{5} + \frac{2\ x\ y}{5} + \frac{3\ x^2}{5}) + e\ m\ y$$

Enter truncation order:
2;

Done with step 1 of 2

Choices for limit cycle amplitude:

1)   0

2) $- 2\ sqrt(5)\ sqrt(- \dfrac{m}{alpha})$

3) $2\ sqrt(5)\ sqrt(- \dfrac{m}{alpha})$

Enter choice number
3;

Done with step 2 of 2

$$x = 2 \ \text{sqrt}(- \frac{m}{alpha}) \ \text{sqrt}(5) \ \cos(z) + \ . \ . \ .$$

$$y = 2 \ \sin(z) \ \text{sqrt}(- \frac{m}{alpha}) \ \text{sqrt}(5) + \ . \ . \ .$$

$$w = 1 - \frac{11 \ e \ m^2}{2}$$

[VAX 8500 TIME = 12 SEC.]

That is, to lowest order approximation, the limit cycle amplitude is calculated as (cf. eq.(18)):

(21) $$(X^2 + Y^2)^{1/2} = 2 \left[- \frac{5M}{\alpha} \right]^{1/2} \quad \text{or} \quad (x^2 + y^2)^{1/2} = 2 \left[- \frac{5\mu}{\alpha} \right]^{1/2}$$

Eq.(21) asserts that for small μ the limit cycle exists only if μ and α have opposite signs. As a check on this computation, we present in Fig.5 the results of a numerical integration of the differential equations (12)-(14) for parameter values $\mu = 0.01$, $\alpha = -0.01$. As can be seen from the Figure, the foregoing result based on center manifold and Lindstedt methods is in good agreement with the numerical computation.

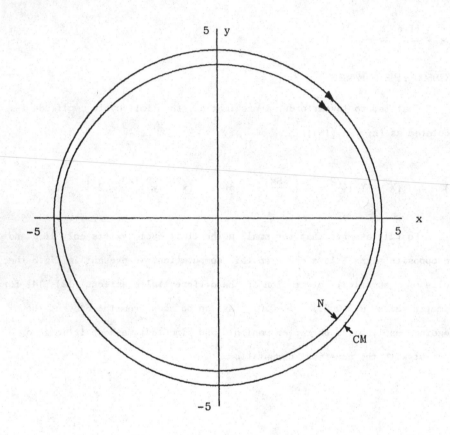

Fig.5. Numerical (= N) and center manifold/ Lindstedt
O(2) (= CM) solutions for the limit cycle of eqs.(12)-(14)
for μ = 0.01 and α = -0.01, displayed in the x-y plane.

Eigencoordinates

The program CM requires that the coordinates on the center manifold be linearly uncoupled from the other coordinates of the problem. In the case that the original system is not in this form, some linear algebra needs to be done. We shall give an example of this complication next, as we consider a bifurcation in the well-known Lorenz equations [14]:

$$(22) \qquad x' = \sigma (y - x)$$

$$(23) \qquad y' = \rho x - y - x z$$

$$(24) \qquad z' = - \beta z + x y$$

We shall be interested in the behavior of this system in the neighborhood of the equilibrium at the origin. Linearizing about the origin, we obtain

$$(25) \qquad \begin{bmatrix} x' \\ y' \\ z' \end{bmatrix} = \begin{bmatrix} -\sigma & \sigma & 0 \\ \rho & -1 & 0 \\ 0 & 0 & -\beta \end{bmatrix} \begin{bmatrix} x \\ y \\ z \end{bmatrix}$$

We use the EIGEN package in MACSYMA to compute the eigenvalues of the linearized system: (Here and in what follows we use the symbols s,r,b and e in MACSYMA to stand for σ, ρ, β and ϵ respectively. We refer to the matrix of the linearized system (25) as A.)

```
A:MATRIX([-S,S,0],[R,-1,0],[0,0,-B]);
                [ - s    s    0 ]
                [               ]
                [  r   - 1    0 ]
                [               ]
                [  0    0   - b ]
```

EIGENVALUES(A);

$$[[-\frac{sqrt(s^2 + (4\ r - 2)\ s + 1) + s + 1}{2},$$

$$\frac{sqrt(s^2 + (4\ r - 2)\ s + 1) - s - 1}{2}, - b], [1, 1, 1]]$$

Here the first list of three eigenvalues is followed by a second list
of their respective multiplicities. Note that at $\rho = 1$ one of the eigenvalues
is zero for all σ and β. Thus at $\rho = 1$ we have a one-dimensional center
manifold. We propose to use our program CM to study the bifurcations of
equilibria in the neighborhood of the origin. Before we can proceed, however,
we must uncouple the eigencoordinate corresponding to the center manifold from
the other eigencoordinates. We return to our previous MACSYMA session and
compute the eigenvectors of the linearized system when $\rho = 1$:

EIGENVECTORS(EV(A,R=1));

$$[[[- s - 1, - b, 0], [1, 1, 1]], [1, - \frac{1}{s}, 0], [0, 0, 1], [1, 1, 0]]$$

Here the first list contains the eigenvalues, the second their
respective multiplicities, and the last three lists are their respective
eigenvectors. In order to diagonalize the linearized system, we transform
variables from x,y,z to u,v,w via a matrix whose columns are the preceding
eigenvectors:

$$(26) \qquad \begin{bmatrix} x \\ y \\ z \end{bmatrix} = \begin{bmatrix} 1 & 1 & 0 \\ 1 & -1/\sigma & 0 \\ 0 & 0 & 1 \end{bmatrix} \begin{bmatrix} u \\ v \\ w \end{bmatrix}$$

Now we need to substitute (26) into (22)-(24). We will use our own
MACSYMA utility program, TRANSFORM, to accomplish this. TRANSFORM performs an

arbitrary coordinate transformation (not necessarily linear) and will be used

again in later Chapters. We follow the MACSYMA run with a listing of

TRANSFORM:

TRANSFORM();

Enter number of equations
3;
Enter symbol for original variable 1
X;
Enter symbol for original variable 2
Y;
Enter symbol for original variable 3
Z;
Enter symbol for transformed variable 1
U;
Enter symbol for transformed variable 2
V;
Enter symbol for transformed variable 3
W;

The RHS's of the d.e.'s are functions of the original variables:
Enter RHS of x d.e.
d x /dt =
S*(Y-X);

d x /dt = s (y - x)

Enter RHS of y d.e.
d y /dt =
R*X-Y-X*Z;

d y /dt = - x z - y + r x

Enter RHS of z d.e.
d z /dt =
-B*Z+X*Y;

d z /dt = x y - b z

The transformation is entered next:

Enter x as a function of the new variables
x =
U+V;

x = v + u

Enter y as a function of the new variables
y =
U-V/S;

$$y = u - \frac{v}{s}$$

Enter z as a function of the new variables
z =
W;

z = w

$$\left[\left[\frac{du}{dt} = -\frac{s\,(u\,w + (1 - r)\,u) + s\,v\,(w - r + 1)}{s + 1}\right.\right.,$$

$$\frac{dv}{dt} = -\frac{s\,((r - 1)\,u - u\,w) + v\,(s\,(-w + r + 1) + s^2 + 1)}{s + 1},$$

$$\left.\left.\frac{dw}{dt} = -\frac{s\,(b\,w - u^2) + v^2 + (u - s\,u)\,v}{s}\right]\right]$$

[VAX 8500 TIME = 1 SEC.]

Here is the listing of TRANSFORM:


```
TRANSFORM():=(

/* input data */

n:read("Enter number of equations"),

for i:1 thru n do

   x[i]:read("Enter symbol for original variable",i),

for i:1 thru n do

   y[i]:read("Enter symbol for transformed variable",i),

print("The RHS's of the d.e.'s are functions of the original variables:"),

for i:1 thru n do (

   print("Enter RHS of",x[i],"d.e."),

   print("d",x[i],"/dt ="),

   f[i]:read(),

   print("d",x[i],"/dt =",f[i])),

print("The transformation is entered next:"),

for i:1 thru n do (

   print("Enter",x[i],"as a function of the new variables"),
```

```
  print(x[i],"="),

  g[i]:read(),

  print(x[i],"=",g[i])),
```
/* do it */
```
for i:1 thru n do depends([x[i],y[i]],t),

for i:1 thru n do eq[i]:diff(x[i],t)=f[i],

trans:makelist(x[i]=g[i],i,1,n),

for i:1 thru n do treq[i]:ev(eq[i],trans,diff),

treqs:makelist(treq[i],i,1,n),

derivs:makelist(diff(y[i],t),i,1,n),

neweqs:solve(treqs,derivs))$
```

In order to observe the bifurcation of equilibria in the center manifold as ρ passes through unity, we set

$$(27) \qquad\qquad \rho = 1 + \epsilon$$

and we embed the system in a 4-dimensional phase space with $\epsilon' = 0$. The following MACSYMA command substitutes (27) into the list NEWEQS which contains the results of TRANSFORM. The right hand sides of the transformed equations are stored in an array RHS, to be conveniently passed on to the program CM:

```
FOR I:1 THRU 3 DO RHS[I]:EV(PART(NEWEQS,1,I,2),R=1+E);

CM();

ENTER NO. OF EQS.
4;
ENTER DIMENSION OF CENTER MANIFOLD
2;

THE D.E.'S MUST BE ARRANGED SO THAT THE FIRST 2 EQS.
REPRESENT THE CENTER MANIFOLD.  I.E. ALL ASSOCIATED
EIGENVALUES ARE ZERO OR HAVE ZERO REAL PARTS.
```

ENTER SYMBOL FOR VARIABLE NO. 1
E;
ENTER SYMBOL FOR VARIABLE NO. 2
U;
ENTER SYMBOL FOR VARIABLE NO. 3
V;
ENTER SYMBOL FOR VARIABLE NO. 4
W;
ENTER ORDER OF TRUNCATION
2;

ENTER RHS OF EQ. 1
D e /DT =
0;

ENTER RHS OF EQ. 2
D u /DT =
RHS[1];

ENTER RHS OF EQ. 3
D v /DT =
RHS[2];

ENTER RHS OF EQ. 4
D w /DT =
RHS[3];

$$\frac{de}{dt} = 0$$

$$\frac{du}{dt} = - \frac{s\,(u\,w - e\,u) + s\,v\,(w - e)}{s + 1}$$

$$\frac{dv}{dt} = - \frac{s\,(e\,u - u\,w) + v\,(s\,(-w + e + 2) + s^2 + 1)}{s + 1}$$

$$\frac{dw}{dt} = - \frac{s\,(b\,w - u^2) + v^2 + (u - s\,u)\,v}{s}$$

CENTER MANIFOLD:

$$[v = - \frac{e\,s\,u}{s^2 + 2\,s + 1}, \; w = \frac{u^2}{b}]$$

FLOW ON THE C.M.:

$$
\left[\frac{de}{dt} = 0, \quad \frac{du}{dt} = - \frac{s\left(\dfrac{u^3}{b} - e\,u\right) - \dfrac{e\,s^2\,u\left(\dfrac{u^2}{b} - e\right)}{s^2 + 2s + 1}}{s + 1}\right]
$$

[VAX 8500 TIME = 13 SEC.]

The resulting expression from CM, stored in the variable CMDE2, can be cleaned up by using the MACSYMA function FACTOR:

FACTOR(CMDE2);

$$
\left[\frac{de}{dt} = 0, \quad \frac{du}{dt} = - \frac{s\,(s^2 - e\,s + 2s + 1)\,u\,(u^2 - b\,e)}{b\,(s + 1)^3}\right]
$$

The last equation gives an approximation for the flow on the center manifold. There are 1 or 3 equilibria, depending on the sign of the product $\beta\,\epsilon$:

(28) $u = 0$ and $u = \pm\,(\beta\,\epsilon)^{1/2}$

Thus there is a pitchfork bifurcation at $\rho = 1$, see Fig.6.

In order to check the center manifold computation, we calculated the center manifold numerically. It is straightforward to do this since any initial condition close to the origin produces a motion which asymptotically approaches the center manifold. In Fig.7 we plot in the x-y plane the numerical results for Lorenz's original parameter values of $\sigma = 10$, $\beta = 8/3$, and for $\rho = 0.9$. Fig.7 also shows the results of the preceding analysis, obtained as follows: We invert the change of coordinates (26) and transform the center manifold equations:

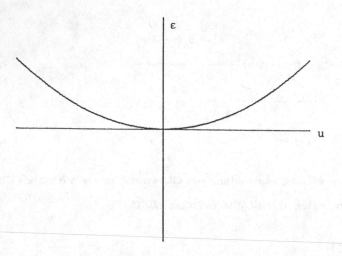

Fig.6. Pitchfork bifurcation of equilibria on the center manifold at the origin in the Lorenz equations, see eq.(28).

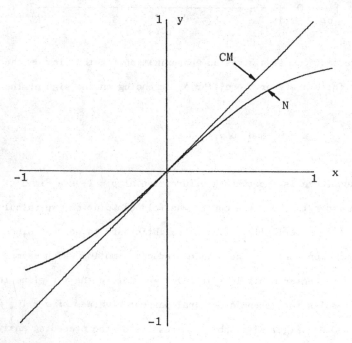

Fig.7. Numerical (= N) and center manifold (= CM) solutions for the center manifold at the origin in the Lorenz equations for $\sigma = 10$, $\beta = 8/3$, $\rho = 0.9$, $\varepsilon = -0.1$, displayed in x-y plane.

(29)
$$v = - \frac{\epsilon \, \sigma \, u}{(1+\sigma)^2} \quad , \quad w = \frac{u^2}{\beta}$$

giving

(30)
$$y = \frac{(\sigma+1)^2 + \epsilon}{(\sigma+1)^2 - \epsilon \, \sigma} \, x \quad , \quad z = \frac{(\sigma+1)^4}{\beta \, [(\sigma+1)^2 - \epsilon \, \sigma]^2} \, x^2$$

For $\sigma = 10$, $\beta = 8/3$ and $\epsilon = -0.1$ this projects onto the x-y plane as the straight line

(31)
$$y = 0.99\ldots \, x$$

Exercise

1. This problem concerns the behavior of van der Pol's equation at infinity [19]. If in van der Pol's equation,

(P1)
$$w'' + w - \epsilon \, (1 - w^2) \, w' = 0$$

we set [27]:

(P2)
$$v = w' \, / \, w \quad \text{and} \quad z = 1 \, / \, w$$

and reparameterize time with $d\tau = w^2 \, dt$, we obtain:

(P3.1)
$$v' = - \epsilon \, v + z^2 \, (\epsilon \, v - v^2 - 1)$$

(P3.2)
$$z' = - v \, z^3$$

where primes now represent derivatives with respect to τ.

The equilibrium at infinity, $v = z = 0$, has eigenvalues zero and $-\epsilon$. There is a center manifold tangent to the z-axis. Use the program CM to obtain an approximation for it to $O(z^{12})$. Note that unlike the treatment of van der Pol's equation by Lindstedt's method, we do not assume ϵ is small here.

Introduction

Like Lindstedt's method, the method of normal forms is used for obtaining approximate solutions to ordinary differential equations. In contrast to Lindstedt's method, however, the method of normal forms does not involve expanding the solution in an infinite series. Rather, the idea is to transform the differential equations themselves into a form which is easily solved. The method involves generating a transformation of coordinates (i.e. dependent variables) in the form of an infinite series, and computing the coefficients of the series so that the resulting transformed differential equations are in a normal (i.e., a simple or canonical) form.

The method may be illustrated by considering a system of two differential equations (although the same process applies to an n dimensional system):

(1.1) $x' = f(x,y)$

(1.2) $y' = g(x,y)$

We assume that the coordinates x and y have been chosen so that the origin $x = y = 0$ is an equilibrium point:

(2) $f(0,0) = g(0,0) = 0$

We Taylor expand f and g about the origin,

(3.1) $x' = a x + b y + F(x,y)$

(3.2) $y' = c x + d y + G(x,y)$

where F and G are strictly nonlinear (i.e., contain terms of degree 2 and
higher).

In the case that F and G contain quadratic terms, we transform from x,y
to u,v coordinates via a near-identity transformation with general quadratic
terms:

(4.1) $x = u + A_{120} u^2 + A_{111} u v + A_{102} v^2$

(4.2) $y = v + A_{220} u^2 + A_{211} u v + A_{202} v^2$

where the A_i are to be determined.

We substitute eqs.(4) into eqs.(3) and neglect terms of order 3 (which
do not influence the quadratic coefficients A_i). The resulting equations are
linear in the derivatives u' and v'. We solve for these and again expand about
u = v = 0, neglecting terms of degree 3 and higher:

(5.1) $u' = a u + b v + C_{120} u^2 + C_{111} u v + C_{102} v^2$

(5.2) $v' = c u + d v + C_{220} u^2 + C_{211} u v + C_{202} v^2$

where the coefficients C_i are known linear functions of the A_i. The linear
terms in (5) are identical to the linear terms in (3) due to the near-identity

nature of the transformation (4).

    At this point the coefficients A_i are chosen so as to put eqs.(5) into
a normal form. If the equilibrium point is hyperbolic (i.e. if all the
eigenvalues have nonzero real parts), then we can always remove all nonlinear
terms [14,§3.3]. However, in some problems (involving resonances or repeated
zero eigenvalues) this is not possible. In such cases the choice of the normal
form is somewhat arbitrary [11].

    Once the coefficients in (4) have been determined, we may extend the
transformation from x,y to u,v coordinates to include cubic terms. (Note that
even if eqs.(3) do not contain cubic terms, the transformation (4) will
generally introduce cubic terms.) Proceeding as before we compute the new
equations on u and v neglecting terms of order 4, and choose the coefficients
of the cubic terms in the near-identity transformation in order to best
simplify the resulting differential equations. Note that the transformation up
to and including cubic terms will not affect the already determined quadratic
terms in the normal form.

    Proceeding in this fashion we can (in principle) generate the desired
transformation to any order of accuracy. Note, however, that the use of
truncated power series can be expected to restrict the applicability of the
method to a neighborhood of the origin.

The Case of a Pair of Imaginary Eigenvalues

As our first example of the method we will consider the system

(6.1) $x' = - y + F(x,y)$

(6.2) $y' = x + G(x,y)$

where F and G are strictly nonlinear in x,y. The problem here is to determine the stability of the equilibrium at the origin. Linearization about the origin is no help, since the origin is a center (cf. Hartman's theorem [14]).

Takens [45] has shown that this important class of systems can be put in the normal form:

(7.1) $r' = B_1 r^3 + B_2 r^5 + B_3 r^7 + \cdots$

(7.2) $\theta' = 1 + D_1 r^2 + D_2 r^4 + D_3 r^6 + \cdots$

where r and θ are polar coordinates

(8) $u = r \cos \theta, \quad v = r \sin \theta$

and where u,v are related to x,y by a near-identity transformation. In rectangular coordinates, eqs.(7) become:

(9.1) $u' = - v + B_1 (u^2 + v^2) u - D_1 (u^2 + v^2) v + 0(5)$

(9.2) $v' = u + B_1 (u^2 + v^2) v + D_1 (u^2 + v^2) u + 0(5)$

Comparison of eqs.(9) and (5) shows that all the quadratic terms can be
eliminated in this case, although terms of odd power will in general remain.

We shall present several MACSYMA programs which will be used to treat
this problem [38]. Before presenting a sample run, we discuss a computational
detail. If the original equations are written in the form

$$(10) \qquad\qquad x' = f(x)$$

where x and f(x) are n-vectors, cf. eqs.(1), then the application of a
near-identity transformation,

$$(11) \qquad\qquad x = u + g(u)$$

where u and g(u) are n-vectors, cf. eqs.(4), gives:

$$(12) \qquad\qquad (I + Dg)\, u' = f(u + g(u))$$

where Dg is the n x n Jacobian matrix. Eq.(12) may be solved for u' as

$$(13) \qquad\qquad u' = (I + Dg)^{-1} f(u + g(u))$$

Now this matrix inversion may be accomplished approximately by using the
expansion:

$$(14) \qquad (I + Dg)^{-1} = I - Dg + (Dg)^{2} - (Dg)^{3} + \cdots + (-Dg)^{m-1}$$

where m is the degree of the highest order terms retained in (13) after
truncation. (The series (14) is truncated at degree m-1 terms since it is
multiplied by f(u+g(u)) which begins with linear terms.) We use this scheme in

the computer algebra program NF which follows. We found it to be faster than

using the MACSYMA SOLVE routine to isolate the derivatives u' in eq.(12).

We return now to eqs.(6) which we write in the truncated form:

$$(15.1) \qquad x' = -y + F_{xx}\frac{x^2}{2} + F_{xy}\,x\,y + F_{yy}\frac{y^2}{2} + F_{xxx}\frac{x^3}{6} + F_{xxy}\frac{x^2 y}{2}$$

$$F_{xyy}\frac{x\,y^2}{2} + F_{yyy}\frac{y^3}{6} + \cdots$$

$$(15.2) \qquad y' = x + G_{xx}\frac{x^2}{2} + G_{xy}\,x\,y + G_{yy}\frac{y^2}{2} + G_{xxx}\frac{x^3}{6} + G_{xxy}\frac{x^2 y}{2}$$

$$G_{xyy}\frac{x\,y^2}{2} + G_{yyy}\frac{y^3}{6} + \cdots$$

The MACSYMA run which applies normal forms to this system consists of

the functions NF, GEN, DECOMPOSE, HOPF2, HOPF3, and a function used in

Chapter 2, TRANSFORM. We will annotate the run in *Italics* (to distinguish the

comments from the run itself) and will follow it with the program listings:

```
NF()$
DO YOU WANT TO ENTER NEW VARIABLE NAMES (Y/N)?
Y;
HOW MANY EQS
2;
SYMBOL FOR OLD X[ 1 ]
X;
SYMBOL FOR OLD X[ 2 ]
Y;
SYMBOL FOR NEW X[ 1 ]
U;
SYMBOL FOR NEW X[ 2 ]
V;
DO YOU WANT TO ENTER NEW D.E.'S (Y/N)?
Y;
```

ENTER RHS OF EQ. NO. 1 ,D x /DT =
-Y+FXX/2*X^2+FXY*X*Y+FYY/2*Y^2+FXXX/6*X^3+FXXY/2*X^2*Y+FXYY/2*X*Y^2+FYYY/6*Y^3;

$$D x /DT = \frac{fyyy \ y^3}{6} + \frac{fxyy \ x \ y^2}{2} + \frac{fyy \ y^2}{2} + \frac{fxxy \ x^2 \ y}{2} + fxy \ x \ y - y + \frac{fxxx \ x^3}{6}$$

$$+ \frac{fxx \ x^2}{2}$$

ENTER RHS OF EQ. NO. 2 ,D y /DT =
X+GXX/2*X^2+GXY*X*Y+GYY/2*Y^2+GXXX/6*X^3+GXXY/2*X^2*Y+GXYY/2*X*Y^2+GYYY/6*Y^3;

$$D y /DT = \frac{gyyy \ y^3}{6} + \frac{gxyy \ x \ y^2}{2} + \frac{gyy \ y^2}{2} + \frac{gxxy \ x^2 \ y}{2} + gxy \ x \ y + \frac{gxxx \ x^3}{6} + \frac{gxx \ x^2}{2}$$

$$+ x$$

INPUT NEAR-IDENTITY TRANSFORMATION
(USE PREV FOR PREVIOUS TRANSFORMATION)
x = u + ?

We wish to enter a quadratic near-identity transformation with general

coefficients, as in eq.(4). The program GEN is designed to facilitate this

frequently needed step.

GEN(2);

$$x = a_{1, \ [0, \ 2]} \ v^2 + a_{1, \ [1, \ 1]} \ u v + a_{1, \ [2, \ 0]} \ u^2 + u$$

y = v + ?
GEN(2);

$$y = a_{2, \ [0, \ 2]} \ v^2 + a_{2, \ [1, \ 1]} \ u v + v + a_{2, \ [2, \ 0]} \ u^2$$

ENTER TRUNCATION ORDER (HIGHEST ORDER TERMS TO BE KEPT)
2;

$$\frac{du}{dt} = - v - ((2 a_{2, \ [2, \ 0]} + 2 a_{1, \ [1, \ 1]} - fxx) \ u^2$$

$$+ (2 a_{2, \ [1, \ 1]} - 4 a_{1, \ [2, \ 0]} + 4 a_{1, \ [0, \ 2]} - 2 fxy) \ v u$$

$$+ (2 a_{2, \ [0, \ 2]} - 2 a_{1, \ [1, \ 1]} - fyy) \ v^2)/2 + \ . \ . \ .$$

$$\frac{dv}{dt} = u - ((2a_{2,[1,1]} - 2a_{1,[2,0]} - gxx) u^2$$

$$+ (-4a_{2,[2,0]} + 4a_{2,[0,2]} - 2a_{1,[1,1]} - 2gxy) v u$$

$$+ (-2a_{2,[1,1]} - 2a_{1,[0,2]} - gyy) v^2)/2 + \ldots$$

DO YOU WANT TO ENTER ANOTHER TRANSFORMATION (Y/N)
N;

[VAX 8500 TIME = 9 SEC.]

These last equations correspond to eqs.(5) in our general treatment. We must now isolate the coefficients C_i (see eqs.(5)), a frequently needed step which is facilitated by the utility program DECOMPOSE. Then we wish to choose the coefficients $a_{i,[j,k]}$ so that the transformed equations are in the normal form (9). The program HOPF2 is designed to automate these steps:

HOPF2();

$$[[a_{1,[2,0]} = -\frac{2 gyy + gxx + 2 fxy}{6},$$

$$a_{2,[2,0]} = \frac{-2 gxy + 2 fyy + fxx}{6}, \quad a_{1,[1,1]} = \frac{gxy - fyy + fxx}{3},$$

$$a_{2,[1,1]} = -\frac{gyy - gxx + fxy}{3}, \quad a_{1,[0,2]} = \frac{-gyy - 2 gxx + 2 fxy}{6},$$

$$a_{2,[0,2]} = \frac{2 gxy + fyy + 2 fxx}{6}]]$$

[VAX 8500 TIME = 6 SEC.]

Having obtained the unknown quadratic coefficents $a_{i,[j,k]}$, we return to NF again to check the preceding work and to extend it to cubic terms:

```
NF()$
DO YOU WANT TO ENTER NEW VARIABLE NAMES (Y/N)?
N;
DO YOU WANT TO ENTER NEW D.E.'S (Y/N)?
N;
INPUT NEAR-IDENTITY TRANSFORMATION
(USE PREV FOR PREVIOUS TRANSFORMATION)
x = u + ?
```

The results of the previous step (which may be referred to by %) may be

substituted into the previous transformation (called PREV here) as follows:

```
EV(PREV,%);
```

$$x = \frac{(- gyy - 2\ gxx + 2\ fxy)\ v^2}{6} + \frac{(gxy - fyy + fxx)\ u\ v}{3}$$

$$- \frac{(2\ gyy + gxx + 2\ fxy)\ u^2}{6} + u$$

```
y = v + ?
EV(PREV,%);
```

$$y = \frac{(2\ gxy + fyy + 2\ fxx)\ v^2}{6} - \frac{(gyy - gxx + fxy)\ u\ v}{3} + v$$

$$+ \frac{(- 2\ gxy + 2\ fyy + fxx)\ u^2}{6}$$

```
ENTER TRUNCATION ORDER (HIGHEST ORDER TERMS TO BE KEPT)
2;

du
-- = - v + . . .
dt

dv
-- = u + . . .
dt
```

This confirms that the routine is working so far. Now we add general cubic

terms to the previous transformation:

DO YOU WANT TO ENTER ANOTHER TRANSFORMATION (Y/N)
Y;
INPUT NEAR-IDENTITY TRANSFORMATION
(USE PREV FOR PREVIOUS TRANSFORMATION)
x = u + ?
PREV+GEN(3);

$$
x = a_{1,\,[0,\,3]}\,v^3 + a_{1,\,[1,\,2]}\,u\,v^2 + \frac{(-\,gyy - 2\,gxx + 2\,fxy)\,v^2}{6}
$$

$$
+\; a_{1,\,[2,\,1]}\,u^2\,v + \frac{(gxy - fyy + fxx)\,u\,v}{3} + a_{1,\,[3,\,0]}\,u^3
$$

$$
-\; \frac{(2\,gyy + gxx + 2\,fxy)\,u^2}{6} + u
$$

y = v + ?
PREV+GEN(3);

$$
y = a_{2,\,[0,\,3]}\,v^3 + a_{2,\,[1,\,2]}\,u\,v^2 + \frac{(2\,gxy + fyy + 2\,fxx)\,v^2}{6}
$$

$$
+\; a_{2,\,[2,\,1]}\,u^2\,v - \frac{(gyy - gxx + fxy)\,u\,v}{3} + v + a_{2,\,[3,\,0]}\,u^3
$$

$$
+\; \frac{(-\,2\,gxy + 2\,fyy + fxx)\,u^2}{6}
$$

ENTER TRUNCATION ORDER (HIGHEST ORDER TERMS TO BE KEPT)
3;

$$
\frac{du}{dt} = -\,v - (((fxy + gxx + 2\,gyy)\,fxx - 2\,fxy\,fyy + 2\,fxy\,gxy + 6\,a_{2,\,[3,\,0]}
$$
      + 6 additional lines omitted here for brevity

$$
\frac{dv}{dt} = u + ((gxy\,fxx + 2\,gxy\,fyy - 2\,gxy^2 - 2\,gxx\,fxy - gxx^2 - 2\,gyy\,gxx
$$

      + 7 additional lines omitted here for brevity

DO YOU WANT TO ENTER ANOTHER TRANSFORMATION (Y/N)
N;

[VAX 8500 TIME = 47 SEC.]

Next we call HOPF3 which DECOMPOSEs the last equations and selects the cubic $a_{i,[j,k]}$ coefficients such that we obtain the normal form (9). Note that this involves 6 linear algebraic equations in 8 unknowns and thus the general solution contains two arbitrary constants. Happily the MACSYMA function SOLVE automatically does the linear algebra for us, and assigns the arbitrary constants the values %r1 and %r2:

HOPF3();

$$[[a_{1,[3,0]} = (2\ gyy^2 + (gxx + 7\ fxy)\ gyy - 3\ gxyy - 2\ gxy^2$$

$$+ (-\ fyy - 7\ fxx)\ gxy - gxxx + 3\ gxx^2 + fxy\ gxx - fyyy - 3\ fyy^2 - fxx\ fyy$$

$$+ 2\ fxy^2 - 3\ fxxy - 2\ fxx^2 + 24\ \%r1)/24,\ a_{2,[3,0]} = \%r2,$$

+ 13 additional lines giving the other $a_{i,[j,k]}$'s in terms of the given constants of the problem and the two arbitrary constants %r1 and %r2.

[VAX 8500 TIME = 34 SEC.]

Next we substitute these values of the $a_{i,[j,k]}$'s into the transformed differential equations, stored in the variable NEWDES by the program NF:

EXPAND(EV(NEWDES,%));

$$[\frac{du}{dt} = \frac{gyy^2\ v^3}{24} + \frac{5\ gxx\ gyy\ v^3}{48} - \frac{5\ fxy\ gyy\ v^3}{48} - \frac{gxyy\ v^3}{16} + \frac{gxy^2\ v^3}{24}$$

+ 48 additional terms, none of which contain either of the arbitrary constants %r1 or %r2

$$\frac{dv}{dt} = \frac{gyyy\ v^3}{16} - \frac{gxy\ gyy\ v^3}{16} + \frac{fyy\ gyy\ v^3}{16} - \frac{gxx\ gxy\ v^3}{16} + \frac{gxxy\ v^3}{16} - \frac{fxx\ gxx\ v^3}{16}$$

+ 47 additional terms with no %r1 or %r2 dependence.

[VAX 8500 TIME = 14 SEC.]

Note that although the transformation to normal form is not unique (since %r1

and %r2 are arbitrary), the normal form itself does not depend on %r1 or %r2.

Next we use the function TRANSFORM given in Chapter 2 to convert the

transformed differential equations to the polar form (7):


```
TRANSFORM();
Enter number of equations
2;
Enter symbol for original variable 1
U;
Enter symbol for original variable 2
V;
Enter symbol for transformed variable 1
R;
Enter symbol for transformed variable 2
THETA;
The RHS's of the d.e.'s are functions of the original variables:
Enter RHS of u d.e.
d u /dt =
```

We specify the previously obtained differential equations by using the MACSYMA

symbol % :


```
RHS(PART(%,1));
Enter RHS of v d.e.
d v /dt =
RHS(PART(%,2));
The transformation is entered next:
Enter u as a function of the new variables
u =
R*COS(THETA);
u = r cos(theta)
Enter v as a function of the new variables
v =
R*SIN(THETA);
v = r sin(theta)
```

$$
\begin{aligned}
[[\frac{dr}{dt} = ((gyyy + (fyy - gxy)\, gyy - gxx\, gxy + gxxy - fxx\, gxx + fxy\, fyy \\
+ fxyy + fxx\, fxy + fxxx)\, r^{3}\, sin^{2}(theta) \\
+ (gyyy + (fyy - gxy)\, gyy - gxx\, gxy + gxxy - fxx\, gxx + fxy\, fyy + fxyy \\
+ fxx\, fxy + fxxx)\, r^{3}\, cos^{2}(theta))/16,
\end{aligned}
$$

$$\frac{dtheta}{dt} = -((2\ gyy^2 + (5\ gxx - 5\ fxy)\ gyy - 3\ gxyy + 2\ gxy^2$$

+ 5 *additional lines omitted for brevity.*

[VAX 8500 TIME = 21 SEC.]

This polar form can be simplified by using the MACSYMA function TRIGSIMP:

TRIGSIMP(%);

$$[[\frac{dr}{dt} = (gyyy + (fyy - gxy)\ gyy - gxx\ gxy + gxxy - fxx\ gxx + fxy\ fyy$$

$$+ fxyy + fxx\ fxy + fxxx)\ r^3 /16,\ \frac{dtheta}{dt} =$$

$$- ((2\ gyy^2 + (5\ gxx - 5\ fxy)\ gyy - 3\ gxyy + 2\ gxy^2 + (-\ fyy - 5\ fxx)\ gxy$$

$$- 3\ gxxx + 5\ gxx^2 - fxy\ gxx + 3\ fyyy + 5\ fyy^2 + 5\ fxx\ fyy + 2\ fxy^2 + 3\ fxxy$$

$$+ 2\ fxx^2)\ r^2 - 48)/48]]$$

[VAX 8500 TIME = 24 SEC.]

Here is the program listing:

```
/* THIS FILE CONTAINS NF(), A NORMAL FORM UTILITY FUNCTION.

   IT ALSO CONTAINS THESE ADDITIONAL FUNCTIONS:

   GEN(N) WILL GENERATE A HOMOGENEOUS ORDER N TRANSFORMATION.

   DECOMPOSE() ISOLATES THE COEFFICIENTS OF THE NEW EQUATIONS.

   VARS(N) GENERATES A LIST OF UNKNOWN COEFFICIENTS OF DEGREE N.

   HOPFk(), FOR k=2,3,4,5,6,7 SOLVES FOR THE COEFFICIENTS OF A SYSTEM OF

   2 DE'S SO AS TO PUT THE EQS IN HOPF NORMAL FORM */
```

```
NF():= BLOCK(

/* NEW VARIABLE NAMES? */

TEST : READ ("DO YOU WANT TO ENTER NEW VARIABLE NAMES (Y/N)?"),

IF TEST = N THEN GO(JUMP),

N : READ ("HOW MANY EQS"),

FOR I THRU N DO (X[I] : READ ("SYMBOL FOR OLD X[",I,"]")),

FOR I THRU N DO( Y[I] : READ ("SYMBOL FOR NEW X[",I,"]")),

FOR I THRU N DO DEPENDS([X[I],Y[I]],T),

KILL(FLAG), /* FLAG USED IN GEN */


JUMP,

/* NEW D.E.'S? */

PRINT ("DO YOU WANT TO ENTER NEW D.E.'S (Y/N)?"),

TEST:READ(),

IF TEST = N THEN GO(LOOP),

FOR I THRU N DO

        (RHS[I]:READ("ENTER RHS OF EQ. NO.",I,",D",X[I],"/DT ="),

          PRINT("D",X[I],"/DT =",RHS[I])),

KILL(RHS2),

RHS2[I,J] := RHS[I],

RHS3:GENMATRIX(RHS2,N,1),


LOOP,

/* NEAR-IDENTITY TRANSFORMATION */

PRINT("INPUT NEAR-IDENTITY TRANSFORMATION

(USE PREV FOR PREVIOUS TRANSFORMATION)"),

FOR I THRU N DO

        (ROW:I,

          PREV :TR[I],
```

```
          TR[I] :READ (X[I],"=",Y[I],"+ ?")。

          PRINT (X[I],"=",Y[I]+TR[I]))。
```

/* INPUT TRUNCATION ORDER */

TRANS : MAKELIST (X[I]=Y[I]+TR[I],I,1,N)。

M : READ("ENTER TRUNCATION ORDER (HIGHEST ORDER TERMS TO BE KEPT)")。

/* TRANSFORM THE D.E.'S */

TEMP2 :EV(RHS3,TRANS)。

/* SOLVE FOR THE TRANSFORMED DERIVATIVES */

KILL(JACOB)。

JACOB[I,J]:=DIFF(TR[I],Y[J])。

JACOB2:GENMATRIX(JACOB,N,N)。

TEMP3:SUM((-1)^I*JACOB2^^I,I,0,M-1).TEMP2。

/* TAYLOR EXPAND THE RESULTING EQS */

NEWRHS : TAYLOR(TEMP3,MAKELIST(Y[I],I,1,N),0,M)。

NEWDES:MAKELIST(DIFF(Y[I],T)=NEWRHS[I,1],I,1,N)。

FOR I:1 THRU N DO

          PRINT(PART(NEWDES,I))。

/* ENTER ANOTHER TRANSFORMATION? */

BRANCH:READ("DO YOU WANT TO ENTER ANOTHER TRANSFORMATION (Y/N)")。

IF BRANCH = Y THEN GO(LOOP)。

NEWDES)$

/* AUXILIARY FUNCTIONS */

GEN(NN):=(

IF NOT NUMBERP(FLAG[NN]) THEN (

          SUB:MAKELIST(K[I],I,1,N)。

          VAR:PRODUCT(Y[I]^K[I],I,1,N)。

          TEMPGEN:A[ROWDUMMY,SUB]*VAR,

          FOR I:1 THRU N DO

```
                TEMPGEN:SUM(EV(TEMPGEN,K[I]=J),J,0,NN),

            TEMPGEN2:LAST(TAYLOR(TEMPGEN,MAKELIST(Y[I],I,1,N),0,NN)),

            TEMPGEN3[NN]:EXPAND(TEMPGEN2),

            FLAG[NN] : 1),

EV(TEMPGEN3[NN],ROWDUMMY=ROW)) $


DECOMPOSE():=(

KILL(C),

IF NOT NUMBERP(FLAG[M]) THEN GEN(M),

TEMP8:SUBST("[","+",TEMPGEN3[M]),

TERMS:EV(TEMP8,A[DUMMY,SUB]:=1),

COEFFS:EV(TEMP8,A[DUMMY,SUB]:=C[DUMMY,SUB],MAKELIST(Y[I]=1,I,1,N)),

FOR I:1 THRU N DO(

    FOR J:1 THRU LENGTH(TERMS) DO(

        EV(PART(COEFFS,J),ROWDUMMY=I)::

        RATCOEF(EXPAND(NEWRHS[I,1]),PART(TERMS,J)))))$


VARS(NN):=(

TEMP5:SUM(EV(TEMPGEN3[NN]),ROWDUMMY,1,N),

TEMP6:SUBST("[","+",TEMP5),

TEMP7:EV(TEMP6,MAKELIST(Y[I]=1,I,1,N)))$


 HOPF2():=(DECOMPOSE(),

        SOLVE([C[1,[2,0]],C[1,[1,1]],C[1,[0,2]],C[2,[2,0]],

            C[2,[1,1]],C[2,[0,2]]],

            VARS(2)))$
```

```
HOPF3():=(DECOMPOSE(),
         SOLVE([C[1,[3,0]]=C[1,[1,2]],C[1,[3,0]]=C[2,[2,1]],
           C[1,[3,0]]=C[2,[0,3]],C[1,[0,3]]=C[1,[2,1]],
           C[1,[0,3]]=-C[2,[3,0]],C[1,[0,3]]=-C[2,[1,2]]],
         VARS(3)))$


HOPF4():=(DECOMPOSE(),
         SOLVE([C[1,[4,0]],C[1,[3,1]],C[1,[2,2]],C[1,[1,3]],
           C[1,[0,4]],C[2,[4,0]],C[2,[3,1]],C[2,[2,2]],
           C[2,[1,3]],C[2,[0,4]]],
         VARS(4)))$


HOPF5():=(DECOMPOSE(),
         SOLVE([C[1,[5,0]]=C[1,[3,2]]/2,C[1,[5,0]]=C[1,[1,4]],
           C[1,[5,0]]=C[2,[4,1]],C[1,[5,0]]=C[2,[2,3]]/2,
           C[1,[5,0]]=C[2,[0,5]],C[2,[5,0]]=C[2,[3,2]]/2,
           C[2,[5,0]]=C[2,[1,4]],C[2,[5,0]]=-C[1,[4,1]],
           C[2,[5,0]]=-C[1,[2,3]]/2,C[2,[5,0]]=-C[1,[0,5]]],
         VARS(5)))$


HOPF6():=(DECOMPOSE(),
         SOLVE([C[1,[6,0]],C[1,[5,1]],C[1,[4,2]],C[1,[3,3]],
           C[1,[2,4]],C[1,[1,5]],C[1,[0,6]],C[2,[6,0]],
           C[2,[5,1]],C[2,[4,2]],C[2,[3,3]],C[2,[2,4]],
           C[2,[1,5]],C[2,[0,6]]],
         VARS(6)))$
```

```
HOPF7():=(DECOMPOSE(),

       SOLVE([C[1,[7,0]]=C[1,[5,2]]/3,C[1,[7,0]]=C[1,[3,4]]/3,

             C[1,[7,0]]=C[1,[1,6]],C[1,[7,0]]=C[2,[6,1]],

             C[1,[7,0]]=C[2,[4,3]]/3,C[1,[7,0]]=C[2,[2,5]]/3,

             C[1,[7,0]]=C[2,[0,7]],C[2,[7,0]]=C[2,[5,2]]/3,

             C[2,[7,0]]=C[2,[3,4]]/3,C[2,[7,0]]=C[2,[1,6]],

             C[2,[7,0]]=-C[1,[6,1]],C[2,[7,0]]=-C[1,[4,3]]/3,

             C[2,[7,0]]=-C[1,[2,5]]/3,C[2,[7,0]]=-C[1,[0,7]]],

       VARS(7)))$
```

Thus we have shown that the problem specified by eqs.(15) can be transformed to the normal form:

$$(16) \qquad r' = B_1 r^3 + \cdots, \qquad \theta' = 1 + D_1 r^2 + \cdots$$

where

$$(17) \qquad 16 B_1 = G_{yyy} + G_{xxy} + F_{xyy} + F_{xxx} + F_{yy} G_{yy} - F_{xx} G_{xx}$$

$$- G_{xx} G_{xy} - G_{yy} G_{xy} + F_{xx} F_{xy} + F_{yy} F_{xy}$$

and where D_1 is known.

The significance of this calculation is now clear: The stability of the origin is given by the sign of B_1. From (16), $B_1 > 0$ (respectively < 0) makes the origin $r = 0$ unstable (respectively stable).

Note that this problem could not have been handled by Lindstedt's method, which gives no information regarding stability.

As an application of these results, we may now resolve an unfinished question raised in connection with eqs.(1)-(11) in Chapter 2. We showed there that the stability of the origin in the system:

(18.1) $x' = y$

(18.2) $y' = -x - xz$

(18.3) $z' = -z + \alpha x^2$

could be reduced via center manifold theory to the simpler question of the stability of the origin in the system:

(19.1) $x' = y$

(19.2) $y' = -x - \dfrac{3}{5} \alpha x^3 + \dfrac{2}{5} \alpha x^2 y - \dfrac{2}{5} \alpha x y^2 + \cdots$

This system will be in the form of eqs.(6) if we interchange x and y. Computation of the quantity B_1 of eq.(17) gives:

(20) $B_1 = \dfrac{\alpha}{20}$

Thus the origin in the center manifold flow (19), and hence in the original three-dimensional flow (18), is stable if $\alpha < 0$ and unstable if $\alpha > 0$. See Fig.8 where this behavior is confirmed by numerical integration of the system (18).

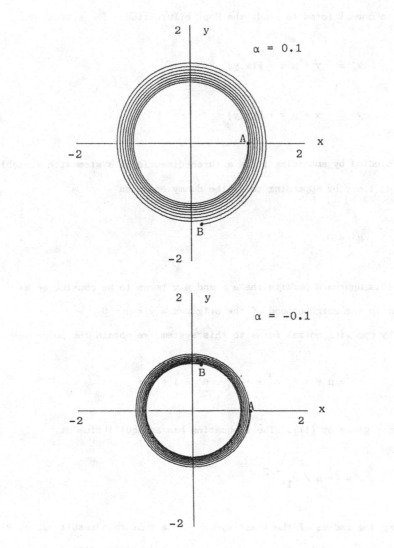

Fig.8. Numerical solutions of the system (18) displayed
in the x-y plane for α = 0.1 and -0.1. In both cases,
A corresponds to the initial condition t=0,x=1,y=0,z=0,
while B corresponds to the system at time t=50.

Before leaving the system (6), we note that by including damping in (6) we may use normal forms to study the Hopf bifurcation. The system

(21.1) $x' = - y + \mu x + F(x,y)$

(21.2) $y' = x + \mu y + G(x,y)$

is best studied by embedding it in a three-dimensional system with variables x,y and μ, i.e., by appending to it the dummy equation:

(21.3) $\mu' = 0$

This approach permits the μx and μy terms to be considered as nonlinear in the neighborhood of the origin $x = y = \mu = 0$.

By applying normal forms to this system, we obtain the polar result:

(22) $r' = \mu r + B_1 r^3 + \cdots \quad , \quad \theta' = 1 + \cdots$

where B_1 is given by (17). The r equation has an equilibrium at

(23) $r = (- \mu / B_1)^{1/2}$

this being the radius of the limit cycle. Note that this result agrees with that derived by Lindstedt's method in Chapter 1, eqs.(26),(27).

Having derived this result by both Lindstedt's method and normal forms, a comparison between the two methods can be made. Most significant is that normal forms provides a <u>dynamic</u> on the process (eq.(22)), while Lindstedt's method offers only the limit cycle radius. Moreover, Lindstedt's method requires the inclusion of a small perturbation parameter ϵ in the equations, while normal forms does not. On the other hand, all results of the method of normal forms are restricted to a neighborhood of the origin, while Lindstedt's method (which is valid in the neighborhood of the Hamitonian system corresponding to $\epsilon = 0$) can be used to approximate limit cycles with large amplitude, i.e. which do not necessarily lie in a neighborhood of the origin. (See, however, Exercise 1.)

The Case of a Double Zero Eigenvalue

In order to motivate our next example of the method of normal forms, we will investigate the behavior of van der Pol's equation at infinity [19],[39]. If in van der Pol's equation,

$$(24) \qquad w'' + w - \epsilon (1 - w^2) w' = 0$$

we set [27]:

$$(25) \qquad x = w / w' \quad \text{and} \quad z = 1 / w'$$

and reparameterize time with $d\tau = w'^2 \, dt$, we obtain:

$$(26.1) \qquad x' = z^2 (1 + x^2 - \epsilon x) + \epsilon x^3$$

$$(26.2) \qquad z' = x z^3 + \epsilon z (x^2 - z^2)$$

where primes now represent derivatives with respect to τ.

As $w' \to \infty$, $z \to 0$, from (25). Moreover x is constant along a radial line in the w,w' plane. Thus as we move outward to infinity along such a line, z approaches 0, and we may think of $z = 0$ as the "line at infinity" parameterized by x. Note that z = 0 satisfies eq.(26.2), and thus is an integral curve for this system.

Substituting ż = 0 into (26.1) shows that x = z = 0 is an equilibrium at infinity for the system (26). We are interested in the nature of the local behavior in the neighborhood of this point.

Note that the system (26) has no linear part. In order to simplify the analysis, we set

$$(27) \qquad\qquad y = z^2$$

whereupon (26) becomes:

$$(28.1) \qquad\qquad x' = y - \epsilon x y + \epsilon x^3 + x^2 y$$

$$(28.2) \qquad\qquad y' = -2 \epsilon y^2 + 2 x y^2 + 2 \epsilon x^2 y$$

The problem here is to determine the stability of the equilibrium at the origin in eqs.(28). Note that linearization about the origin is no help since (28) has a double zero eigenvalue at the origin (cf. Hartman's theorem, [14]).

System (28) is a particular case of the system:

$$(29.1) \qquad\qquad x' = y + F(x,y)$$

$$(29.2) \qquad\qquad y' = G(x,y)$$

where F and G are strictly nonlinear in x,y. Takens [45] has shown that this class of systems can be put in the normal form:

(30.1)
$$u' = v + \sum_{n=2} b_n u^n$$

(30.2)
$$v' = \sum_{n=2} a_n u^n$$

where u and v are related to x and y by a near-identity transformation.

We note that Bogdanov [5] and Guckenheimer and Holmes [14] have chosen the alternate normal form for this problem:

(31.1)
$$U' = V$$

(31.2)
$$V' = \sum_{n=2} a_n U^n + V \sum_{n=2} n b_n U^{n-1}$$

which is obtainable from (30) via the near-identity transformation:

(32)
$$u = U \quad \text{and} \quad v = V - \sum_{n=2} b_n U^n$$

We write eqs.(29) in the truncated form:

(33.1)
$$x' = y + F_{xx} \frac{x^2}{2} + F_{xy} \, x y + F_{yy} \frac{y^2}{2} + F_{xxx} \frac{x^3}{6} + F_{xxy} \frac{x^2 y}{2}$$
$$+ F_{xyy} \frac{x y^2}{2} + F_{yyy} \frac{y^3}{6} + \cdots$$

(33.2)
$$y' = G_{xx} \frac{x^2}{2} + G_{xy} \, x y + G_{yy} \frac{y^2}{2} + G_{xxx} \frac{x^3}{6} + G_{xxy} \frac{x^2 y}{2}$$
$$+ G_{xyy} \frac{x y^2}{2} + G_{yyy} \frac{y^3}{6} + \cdots$$

The following MACSYMA run applies our normal form utilities to this
problem. Our goal is to choose a near-identity transformation which puts the
system (33) into the normal form (30):


```
NF()$
DO YOU WANT TO ENTER NEW VARIABLE NAMES (Y/N)?
Y;
HOW MANY EQS
2;
SYMBOL FOR OLD X[ 1 ]
X;
SYMBOL FOR OLD X[ 2 ]
Y;
SYMBOL FOR NEW X[ 1 ]
U;
SYMBOL FOR NEW X[ 2 ]
V;
DO YOU WANT TO ENTER NEW D.E.'S (Y/N)?
Y;
ENTER RHS OF EQ. NO. 1 ,D x /DT =
Y+FXX*X^2/2+FXY*X*Y+FYY*Y^2/2+FXXX*X^3/6+FXXY*X^2*Y/2+FXYY*X*Y^2/2+FYYY*Y^3/6;
```

$$D \; x \; /DT = \frac{fyyy \; y}{6} + \frac{fxyy \; x \; y}{2} + \frac{fyy \; y}{2} + \frac{fxxy \; x \; y}{2} + fxy \; x \; y + y + \frac{fxxx \; x}{6}$$

$$+ \frac{fxx \; x}{2}$$

```
ENTER RHS OF EQ. NO. 2 ,D y /DT =
GXX*X^2/2+GXY*X*Y+GYY*Y^2/2+GXXX*X^3/6+GXXY*X^2*Y/2+GXYY*X*Y^2/2+GYYY*Y^3/6;
```

$$D \; y \; /DT = \frac{gyyy \; y}{6} + \frac{gxyy \; x \; y}{2} + \frac{gyy \; y}{2} + \frac{gxxy \; x \; y}{2} + gxy \; x \; y + \frac{gxxx \; x}{6} + \frac{gxx \; x}{2}$$

```
INPUT NEAR-IDENTITY TRANSFORMATION
(USE PREV FOR PREVIOUS TRANSFORMATION)
x = u + ?
```


*As in the previous run, the utility program GEN is used to generate a quadratic
near-identity transformation with general coefficients:*

GEN(2);

$$x = a_{1, [0, 2]} v^2 + a_{1, [1, 1]} u v + a_{1, [2, 0]} u^2 + u$$

y = v + ?
GEN(2);

$$y = a_{2, [0, 2]} v^2 + a_{2, [1, 1]} u v + v + a_{2, [2, 0]} u^2$$

ENTER TRUNCATION ORDER (HIGHEST ORDER TERMS TO BE KEPT)
2;

$$\frac{du}{dt} = v + ((2 a_{2, [2, 0]} + fxx) u^2 + (2 a_{2, [1, 1]} - 4 a_{1, [2, 0]} + 2 fxy) v u$$

$$+ (2 a_{2, [0, 2]} - 2 a_{1, [1, 1]} + fyy) v^2)/2 + \ldots$$

$$\frac{dv}{dt} = \frac{gxx \, u^2 + (- 4 a_{2, [2, 0]} + 2 gxy) v u + (- 2 a_{2, [1, 1]} + gyy) v^2}{2} + \ldots$$

DO YOU WANT TO ENTER ANOTHER TRANSFORMATION (Y/N)
N;

[VAX 8500 TIME = 11 SEC.]

Now we use the utility program DECOMPOSE to isolate the coefficients in the
transformed equations. $C_{i, [j, k]}$ *is the coefficient of* $u^j v^k$ *in the* i^{th}
equation:

DECOMPOSE()$

[VAX 8500 TIME = 4 SEC.]

We use the MACSYMA function SOLVE to eliminate the uv *and* v^2 *terms in both*
equations by appropriately choosing the 6 constants $a_{i, [j, k]}$. *The utility*
program VARS is used to generate a list of the unknowns $a_{i, [j, k]}$. *Since there*
are 4 algebraic equations in 6 unknowns, MACSYMA returns the general solution
in terms of 2 arbitrary constants %r1 and %r2:

SOLVE([C[1,[1,1]],C[1,[0,2]],C[2,[1,1]],C[2,[0,2]]],VARS(2));

$$
[[a_{1,\,[2,\,0]} = \frac{gyy + 2\,fxy}{4},\ a_{2,\,[2,\,0]} = \frac{gxy}{2},\ a_{1,\,[1,\,1]} = \frac{fyy + 2\,\%r1}{2},
$$

$$
a_{2,\,[1,\,1]} = \frac{gyy}{2},\ a_{1,\,[0,\,2]} = \%r2,\ a_{2,\,[0,\,2]} = \%r1]]
$$

[VAX 8500 TIME = 1 SEC.]

Next we substitute these values into the transformation by again calling NF.

We refer to the previous transformation as PREV, and the $a_{i,\,[j,k]}$ values as % :

NF()$
DO YOU WANT TO ENTER NEW VARIABLE NAMES (Y/N)?
N;
DO YOU WANT TO ENTER NEW D.E.'S (Y/N)?
N;
INPUT NEAR-IDENTITY TRANSFORMATION
(USE PREV FOR PREVIOUS TRANSFORMATION)
x = u + ?
EV(PREV,%);

$$
x = \%r2\,v^2 + \frac{(fyy + 2\,\%r1)\,u\,v}{2} + \frac{(gyy + 2\,fxy)\,u^2}{4} + u
$$

y = v + ?
EV(PREV,%);

$$
y = \%r1\,v^2 + \frac{gyy\,u\,v}{2} + v + \frac{gxy\,u^2}{2}
$$

ENTER TRUNCATION ORDER (HIGHEST ORDER TERMS TO BE KEPT)
2;

$$
\frac{du}{dt} = v + \frac{(gxy + fxx)\,u^2}{2} + \ldots
$$

$$
\frac{dv}{dt} = \frac{gxx\,u^2}{2} + \ldots
$$

Note that the transformed equations do not depend upon the arbitrary constants
%r1, %r2. Having accomplished the normal form computation up to quadratic
terms, we continue the process in order to include cubic terms:

DO YOU WANT TO ENTER ANOTHER TRANSFORMATION (Y/N)
Y;
INPUT NEAR-IDENTITY TRANSFORMATION
(USE PREV FOR PREVIOUS TRANSFORMATION)
x = u + ?
PREV+GEN(3);

$$x = a_{1,\,[0,\,3]}\, v^3 + a_{1,\,[1,\,2]}\, u\, v^2 + \%r2\, v^2 + a_{1,\,[2,\,1]}\, u^2\, v$$

$$+ \frac{(fyy + 2\,\%r1)\, u\, v}{2} + a_{1,\,[3,\,0]}\, u^3 + \frac{(gyy + 2\, fxy)\, u^2}{4} + u$$

y = v + ?
PREV+GEN(3);

$$y = a_{2,\,[0,\,3]}\, v^3 + a_{2,\,[1,\,2]}\, u\, v^2 + \%r1\, v^2 + a_{2,\,[2,\,1]}\, u^2\, v + \frac{gyy\, u\, v}{2} + v$$

$$+ a_{2,\,[3,\,0]}\, u^3 + \frac{gxy\, u^2}{2}$$

ENTER TRUNCATION ORDER (HIGHEST ORDER TERMS TO BE KEPT)
3;

$$\frac{du}{dt} = v + \frac{(gxy + fxx)\, u^2}{2} - ((3\, gyy\, gxy - 12\, a_{2,\,[3,\,0]} + 6\, gxx\,\%r1 + 3\, gxx\, fyy$$

$$- 2\, fxxx)\, u^3 + ((6\,\%r1 - 3\, fyy)\, gxy - 9\, fxy\, gyy - 12\, a_{2,\,[2,\,1]} - 6\, fxx\,\%r1$$

$$- 6\, fxy^2 + 36\, a_{1,\,[3,\,0]} - 3\, fxx\, fyy + 12\, gxx\,\%r2 - 6\, fxxy)\, v\, u^2$$

$$+ (- 6\, fyy\, gyy - 24\, fxy\,\%r1 - 12\, a_{2,\,[1,\,2]} - 6\, fyy\, fxy + 24\, a_{1,\,[2,\,1]}$$

$$- 12\, fxx\,\%r2 - 6\, fxyy)\, v^2\, u + (- 12\, fyy\,\%r1 - 12\, a_{2,\,[0,\,3]} - 12\,\%r2\, fxy$$

$$+ 12\, a_{1,\,[1,\,2]} - 2\, fyyy)\, v^3)/12 + \ldots$$

$$\frac{dv}{dt} = \frac{gxx\ u^2}{2} - ((6\ fxx\ gxy - 6\ gxx\ fxy - 2\ gxxx)\ u^3$$

$$+ ((- 12\ gyy - 6\ fxy)\ gxy + 36\ a_{2,\ [3,\ 0]} + 3\ fxx\ gyy - 6\ gxx\ fyy - 6\ gxxy)\ v$$

$$u^2 + ((- 24\ \%r1 - 6\ fyy)\ gxy - 6\ gyy^2 ¢ 24\ a_{2,\ [2,\ 1]} - 12\ gxx\ \%r2 - 6\ gxyy)$$

$$v^2\ u + (- 12\ \%r2\ gxy - 12\ \%r1\ gyy + 12\ a_{2,\ [1,\ 2]} - 2\ gyyy)\ v^3\ )/12 + \ .\ .\ .$$

DO YOU WANT TO ENTER ANOTHER TRANSFORMATION (Y/N)
N;

[VAX 8500 TIME = 38 SEC.]

DECOMPOSE()$

[VAX 8500 TIME = 17 SEC.]

*Once again we use DECOMPOSE and SOLVE to obtain the $a_{i,[j,k]}$'s. The general
solution again involves two arbitrary constants, %r3 and %r4:*

SOLVE([C[1,[2,1]],C[1,[1,2]],C[1,[0,3]],C[2,[2,1]],C[2,[1,2]],C[2,[0,3]]],
      VARS(3));

$$[[a_{1,\ [3,\ 0]} = (gyy^2 + 3\ fxy\ gyy + gxyy + (2\ fyy + 2\ \%r1)\ gxy - 2\ \%r2\ gxx$$

$$+ fxx\ fyy + 2\ fxy^2 + 2\ fxxy + 2\ \%r1\ fxx)/12,$$

$$a_{2,\ [3,\ 0]} = \frac{(4\ gxy - fxx)\ gyy + 2\ fxy\ gxy + 2\ gxxy + 2\ fyy\ gxx}{12},$$

$$a_{1,\ [2,\ 1]} = (gyyy + (3\ fyy + 6\ \%r1)\ gyy + 6\ \%r2\ gxy + 3\ fxy\ fyy + 3\ fxyy$$

$$+ 12\ \%r1\ fxy + 6\ \%r2\ fxx)/12,\ a_{2,\ [2,\ 1]} =$$

$$\frac{gyy^2 + gxyy + (fyy + 4\ \%r1)\ gxy + 2\ \%r2\ gxx}{4},$$

$$a_{1, [1, 2]} = \frac{fyyy + 6\ \%r1\ fyy + 6\ \%r2\ fxy + 6\ \%r3}{6},$$

$$a_{2, [1, 2]} = \frac{gyyy + 6\ \%r1\ gyy + 6\ \%r2\ gxy}{6},\ a_{1, [0, 3]} = \%r4,\ a_{2, [0, 3]} = \%r3]]$$

[VAX 8500 TIME = 4 SEC.]

We terminate the calculation by substituting these values into the transformed differential equations, which are stored in the variable NEWDES by NF:

EXPAND(EV(NEWDES,%));

$$\left[\frac{du}{dt} = v + \frac{gxy\ gyy\ u^3}{12} - \frac{fxx\ gyy\ u^3}{12} + \frac{fxy\ gxy\ u^3}{6} + \frac{gxxy\ u^3}{6} - \frac{fyy\ gxx\ u^3}{12}\right.$$

$$- \frac{\%r1\ gxx\ u^3}{2} + \frac{fxxx\ u^3}{6} + \frac{gxy\ u^2}{2} + \frac{fxx\ u^2}{2},$$

$$\left.\frac{dv}{dt} = - \frac{fxx\ gxy\ u^3}{2} + \frac{gxxx\ u^3}{6} + \frac{fxy\ gxx\ u^3}{2} + \frac{gxx\ u^2}{2}\right]$$

[VAX 8500 TIME = 2 SEC.]

Note that this time the normal form does depend upon the arbitrary constant %r1. In the case that the coeffcient gxx is not zero, %r1 can be selected so as to eliminate the u^3 term in the u' equation.

Let us apply the foregoing results to eqs.(28) which represent van der Pol's equation at infinity. In this case

(34)

$$F_{xx} = 0,\ F_{xy} = -\epsilon,\ F_{yy} = 0,\ F_{xxx} = 6\epsilon,\ F_{xxy} = 2,\ F_{xyy} = 0,\ F_{yyy} = 0,$$

$$G_{xx} = 0,\ G_{xy} = 0,\ G_{yy} = -4\epsilon,\ G_{xxx} = 0,\ G_{xxy} = 4\epsilon,\ G_{xyy} = 4,\ G_{yyy} = 0,$$

When these values are substituted into the results of our computer algebra work, we find:

$$(35.1) \qquad u' = v + \frac{5}{3} \epsilon u^3 + O(4)$$

$$(35.2) \qquad v' = 0 + O(4)$$

Determinacy

Now the question is, have we taken enough terms of the series to correctly describe the behavior in the neightborhood of the equilibrium point ? A truncated system is said to be determined if the inclusion of any higher order terms cannot effect the topological nature of the local behavior about the singularity. For systems of the form (30), Takens [45] has shown that if the coefficient a_2 does not vanish, then the flow of the system (30) is topologically equivalent to the flow of the simplified system

$$(36.1) \qquad u' = v$$

$$(36.2) \qquad v' = a_2 u^2$$

This result is inapplicable to eqs.(35), however, since $a_2 = 0$ there.

Takens' determinacy result (36) was extended by Rand & Keith [38] who used MACSYMA to perform a series of "blow-up" transformations necessary to draw the following conclusions:

Let a_n be the first nonzero coefficient in eq.(30.2). The determinacy results fall into two cases, depending upon whether n is even or odd:

   n even: If all the coefficients b_i in eq. (30.1) are zero for
i < (n+2)/2,

(37) $b_2 = b_3 = b_4 = \cdots = b_{n/2} = 0$,

then in some neighborhood of the origin, the flow given by eqs. (30) is
topologically equivalent to the flow given by the simplified system:

(38.1) $u' = v$

(38.2) $v' = a_n u^n$

This result is a natural extension of Takens' result (36). The case where n is
odd is somewhat different, however:

      n odd: If all the coefficients b_i in eq. (30.1) are zero for
i < (n+1)/2,

(39) $b_2 = b_3 = b_4 = \cdots = b_{(n-1)/2} = 0$.

and if, in addition,

(40) $4 a_n + \frac{n+1}{2} b_{\frac{n+1}{2}}^2 > 0$,

then in some neighborhood of the origin, the flow given by eqs. (30) is
topologically equivalent to the flow given by the simplified system:

(41.1) $u' = v + b_{\frac{n+1}{2}} u^{\frac{n+1}{2}}$

(41.2) $v' = a_n u^n$

In order to apply these results to eqs.(28), we must extend the computer algebra treatment beyond eqs.(35) to include terms of higher order, until the first a_n coefficient is nonzero. We omit the MACSYMA run since the procedure is to use the functions NF, DECOMPOSE, and SOLVE just as in the previous run. The results are as follows:

If in eqs.(28) we perform the near-identity transformation

$$(42.1) \qquad x = u - \frac{3}{2}\epsilon\, u^2 + \frac{4+15\epsilon^2}{6}\, u^3 - \frac{70\epsilon+105\epsilon^3}{24}\, u^4 + \frac{88+1086\epsilon^2+945\epsilon^4}{120}\, u^5$$

$$(42.2) \qquad y = v - 2\epsilon uv + \frac{2}{3}\epsilon\, u^3 + (1+4\epsilon^2)\, u^2 v - \frac{7}{3}\epsilon^2\, u^4 - (5\epsilon+8\epsilon^3)\, u^3 v$$

$$+ \frac{24\epsilon+185\epsilon^3}{30}\, u^5 + \frac{16+207\epsilon^2+192\epsilon^4}{12}\, u^4 v$$

we obtain the normal form:

$$(43.1) \qquad u' = v + \frac{5}{3}\epsilon\, u^3 - \frac{5}{2}\epsilon^2\, u^4 + \frac{8\epsilon + 225\,\epsilon^3}{60}\, u^5 + O(6)$$

$$(43.2) \qquad v' = -2\,\epsilon^2\, u^5 + O(6)$$

Eqs.(43) are of the form (30) with $a_5 = -2\epsilon^2$ and $b_3 = 5\epsilon/3$. For these values, $n = 5$ is odd and eq.(40) is satisfied:

$$(44) \qquad 4\, a_5 + 3\, b_3^2 = \frac{\epsilon^2}{3} > 0,$$

and so we may conclude that in the neighborhood of the origin, the flow of (43) is topologically equivalent to the flow of the simplified system:

(45.1) $\qquad u' = v + \dfrac{5}{3} \epsilon\, u^3$

(45.2) $\qquad v' = - 2\, \epsilon^2 u^5$

Eqs.(45) may be simplified by setting $q = u^3$ and reparameterizing time with $dT = u^2\, d\tau$ (recall τ was defined in connection with eqs.(26)). This gives:

(46.1) $\qquad q' = 5\, \epsilon\, q + 3\, v$

(46.2) $\qquad v' = - 2\, \epsilon^2 q$

where primes represent differentiation with respect to T. This linear system has eigenvalues 2ϵ and 3ϵ, and thus is an unstable node for $\epsilon > 0$ (and a stable node for $\epsilon < 0$). In view of the preceding results on determinacy, we may conclude that eqs.(28) are also unstable for $\epsilon > 0$ (and stable for $\epsilon < 0$). See Fig.9 where this behavior is confirmed by numerical integration of eqs.(28). Note that it is sufficient to show that eqs.(28) are unstable for $\epsilon > 0$ since these equations are unchanged when ϵ, t and x are respectively replaced by $-\epsilon$, $-t$ and $-x$, and thus an unstable equilibrium point becomes stable when ϵ changes sign.

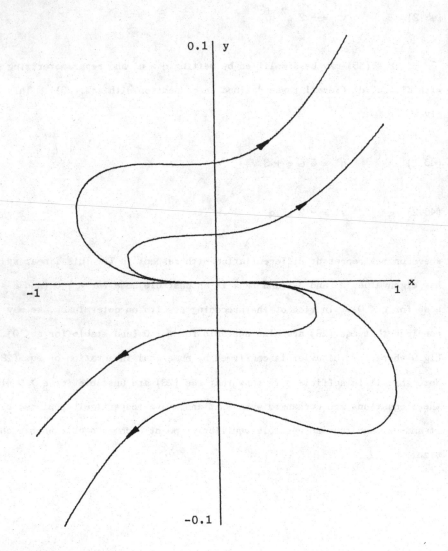

Fig.9. Numerical solution of eqs.(28) for $\varepsilon = 0.1$.

Exercises

1. We have pointed out that the method of normal forms is valid locally in the neighborhood of the equilibrium point. Thus it is not directly applicable to finding the limit cycle in van der Pol's equation,

(P1) $w'' + w + \epsilon (w^2 -1) w' = 0$

since we have seen that Lindstedt's method gave the approximate amplitude of the limit cycle for small ϵ to be 2.

Nevertheless, normal forms can be used to investigate eq.(P1) by using the singular transformation

(P2) $w = \dfrac{x}{\sqrt{\epsilon}}$

which gives

(P3) $x'' + x - \epsilon x' + x^2 x' = 0$

which may be written in the form

(P4.1) $x' = y$

(P4.2) $y' = - x + \epsilon y - x^2 y$

The system (P4) exhibits a Hopf bifurcation at the origin when $\epsilon = 0$.

Show that the linear part of eqs.(P4) may be put in the canonical form

(P5.1) $u' = - \Omega v + \dfrac{\epsilon}{2} u$

(P5.2) $v' = \Omega u + \dfrac{\epsilon}{2} v - u^2 v + \dfrac{\epsilon}{2\Omega} u^3$

by the transformation (see Chapter 1, eq.(29))

(P6.1) $x = u$

(P6.2) $y = \frac{\epsilon}{2} u - \Omega v$

where $\Omega = \sqrt{1 - \frac{\epsilon^2}{4}}$

   Then show that by adding to eqs.(P5) the equation

(P7) $\epsilon' = 0$

a normal forms transformation gives the polar form:

(P8.1) $r' = \frac{\epsilon}{2} r - \frac{1}{8} r^3 + \cdots$

(P8.2) $\theta' = \Omega + \cdots$

   Note that eq.(P8.1) yields the approximate amplitude of the limit cycle for small ϵ to be $2\sqrt{\epsilon}$, which agrees with Lindstedt's method in view of (P2).

2. Show that the following system of two coupled van der Pol oscillators can be treated by the method of normal forms:

(P9.1) $w_1'' + w_1 + \epsilon (w_1^2 - 1) w_1' = A \epsilon w_1 + B \epsilon w_2$

(P9.2) $w_2'' + w_2 + \epsilon (w_2^2 - 1) w_2' = C \epsilon w_1 + D \epsilon w_2$

Begin, as in the previous problem, by setting

(P10)
$$w_i = \frac{x_i}{\sqrt{\epsilon}}$$

Then set $y_i = x_i'$ to obtain a first order system. Following eqs.(P6), set

(P11)
$$x_i = u_i, \quad y_i = \frac{\epsilon}{2} u_i - \Omega v_i, \quad \Omega = \sqrt{1 - \frac{\epsilon^2}{4}}$$

Now use the method of normal forms to show that the near-identity

transformation

(P12.1)
$$u_1 = \xi_1 + \frac{A}{4} \tilde{\epsilon} \, \xi_1 + \frac{B}{4} \tilde{\epsilon} \, \xi_2 + \cdots$$

(P12.2)
$$v_1 = \eta_1 - \frac{A}{4} \tilde{\epsilon} \, \eta_1 - \frac{B}{4} \tilde{\epsilon} \, \eta_2 + \cdots$$

(P12.3)
$$u_2 = \xi_2 + \frac{D}{4} \tilde{\epsilon} \, \xi_2 + \frac{C}{4} \tilde{\epsilon} \, \xi_1 + \cdots$$

(P12.4)
$$v_2 = \eta_2 - \frac{D}{4} \tilde{\epsilon} \, \eta_2 - \frac{C}{4} \tilde{\epsilon} \, \eta_1 + \cdots$$

(P12.5)
$$\epsilon = \tilde{\epsilon}$$

gives, in polar coordinates defined by $\xi_i = r_i \cos \theta_i$, $\eta_i = r_i \sin \theta_i$,
the approximate system:

(P13.1) $\qquad r_1{}' = \frac{\epsilon}{2} r_1 - \frac{r_1{}^3}{8} + \frac{B}{2} \epsilon\, r_2 \sin(\theta_2 - \theta_1) + \cdots$

(P13.2) $\qquad r_2{}' = \frac{\epsilon}{2} r_2 - \frac{r_2{}^3}{8} + \frac{C}{2} \epsilon\, r_1 \sin(\theta_1 - \theta_2) + \cdots$

(P13.3) $\qquad \theta_1{}' = 1 - \frac{A\epsilon}{2} - \frac{B\epsilon}{2} \frac{r_2}{r_1} \cos(\theta_2 - \theta_1) + \cdots$

(P13.4) $\qquad \theta_2{}' = 1 - \frac{D\epsilon}{2} - \frac{C\epsilon}{2} \frac{r_1}{r_2} \cos(\theta_1 - \theta_2) + \cdots$

In deriving eqs.(P13), work to order 3 but neglect terms of $O(\epsilon^2)$. Eqs.(P13) were originally derived by the two variable expansion method [37] and have been studied in [9],[10].

CHAPTER 4

THE TWO VARIABLE EXPANSION METHOD

Introduction

The two variable expansion method [20], also known as the method of multiple scales [32], is an extension of Lindstedt's method. While Lindstedt's method is useful for approximating <u>periodic solutions</u> of o.d.e.'s, it offers no information about the stability of the solutions. The two variable method is used to construct an approximate <u>general solution</u> to a system of o.d.e.'s. It can be used to obtain the behavior of the system in the neighborhood of a limit cycle or an equilibrium point.

In order to fix our ideas, let us again think in terms of van der Pol's equation as an example:

(1)
$$\frac{d^2x}{dt^2} + x = \epsilon \, (1 - x^2) \, \frac{dx}{dt}$$

The method is based on the idea that the approach to a limit cycle occurs on a slower time scale than the time scale of the limit cycle itself. For example, if the motion around the limit cycle has frequency of order unity, then the approach to the limit cycle may occur on a time scale of ϵt. In particular, this is the case for Hopf bifurcations and perturbations off of Hamiltonian systems. Algebraically this means that those terms in the solution which represent the approach to steady state involve an explicit dependence on time t in the form ϵt. The method utilizes this dependence by assuming that the solution is explicitly a function of two time variables, which we shall call ξ and η. Here ξ will represent ordinary time t and η will represent slow time ϵt:

$$(2) \qquad\qquad \xi = t, \quad \eta = \epsilon t$$

The method assumes that the solution x explicitly depends upon both time variables:

$$(3) \qquad\qquad x = x(\xi, \eta)$$

Substitution of (3) into (1) requires the use of the chain rule:

$$(4) \qquad\qquad \frac{dx}{dt} = \frac{\partial x}{\partial \xi} \frac{d\xi}{dt} + \frac{\partial x}{\partial \eta} \frac{d\eta}{dt} = \frac{\partial x}{\partial \xi} + \epsilon \frac{\partial x}{\partial \eta}$$

$$(5) \qquad\qquad \frac{d^2 x}{dt^2} = \frac{\partial^2 x}{\partial \xi^2} + 2 \epsilon \frac{\partial^2 x}{\partial \xi \partial \eta} + \epsilon^2 \frac{\partial^2 x}{\partial \eta^2}$$

In addition to the change of independent variables given by (4) and (5), we expand x in a power series in ϵ:

$$(6) \qquad\qquad x(\xi, \eta) = x_0(\xi, \eta) + \epsilon x_1(\xi, \eta) + \cdots$$

We note at this point that we will use this method only up to terms of order ϵ. Although the method can in principle be extended to arbitrary order in ϵ, we have found that its use beyond the lowest order terms in ϵ is awkward. If one wanted to apply the method to higher order terms, then additional dependence on time could be built into the method by either taking new independent variables such as $\epsilon^2 t$, $\epsilon^3 t$, ..., or by stretching ξ or η, e.g., with $\xi = (1 + \omega_1\epsilon + \omega_2\epsilon^2 + \cdots)$ t. See [32],[20].

Substituting (4)-(6) into van der Pol's eq.(1) and collecting terms, we obtain:

(7) $\qquad \epsilon^0: \quad \dfrac{\partial^2 x_0}{\partial\xi^2} + x_0 = 0$

(8) $\qquad \epsilon^1: \quad \dfrac{\partial^2 x_1}{\partial\xi^2} + x_1 = -2\,\dfrac{\partial^2 x_0}{\partial\xi\partial\eta} + (1 - x_0^2)\,\dfrac{\partial x_0}{\partial\xi}$

Eq.(7) is a partial differential equation. Its general solution involves two arbitrary functions of η:

(9) $\qquad x_0 = A(\eta)\cos\xi + B(\eta)\sin\xi$

Substituting (9) into (8) and simplifying the trig terms gives:

(10) $\qquad \dfrac{\partial^2 x_1}{\partial\xi^2} + x_1 = [\;2A' - A + \dfrac{A}{4}(A^2 + B^2)]\sin\xi$

$$+ [-2B' + B - \dfrac{B}{4}(A^2 + B^2)]\cos\xi$$

$$+ \dfrac{A}{4}(A^2 - 3B^2)\sin 3\xi + \dfrac{B}{4}(B^2 - 3A^2)\cos 3\xi$$

where primes represent differentiation with respect to η.

Elimination of secular terms in (10) is accomplished by requiring the coefficients of $\cos\xi$ and $\sin\xi$ to vanish. This gives a pair of coupled differential equations on $A(\eta)$ and $B(\eta)$ called the <u>slow flow</u> equations:

(11.1)
$$A' = \frac{A}{2} - \frac{A}{8}(A^2 + B^2)$$

(11.2)
$$B' = \frac{B}{2} - \frac{B}{8}(A^2 + B^2)$$

Eqs.(11) are most easily treated by transforming to polar coordinates,

(12)
$$A = R \cos \theta, \quad B = R \sin \theta$$

whereupon we obtain:

(13.1)
$$R' = \frac{R}{2} - \frac{R^3}{8}$$

(13.2)
$$\theta' = 0$$

Now eq.(13.1) is a separable first order o.d.e. It is easily solved:

(14)
$$R(\eta) = \frac{2\,R(0)\,e^{\eta/2}}{\sqrt{(e^\eta - 1)\,R(0)^2 + 4}}$$

This eq. has the property that as $\eta \to \infty$, $R \to 2$, the limit cycle amplitude. As $\eta \to -\infty$, $R \to 0$ if $R(0) < 2$, so that if we run time backwards, all points inside the limit cycle approach the equilibrium at the origin. If $R(0) > 2$, however, the denominator has a zero at $\eta = \log[1 - 4/R(0)^2] < 0$. Thus if we run time backwards, motions starting outside the limit cycle are predicted to escape to infinity in finite time, in agreement with numerical investigations.

In terms of R and θ, eq.(9) for x_0 becomes

(15)
$$x_0 = R(\eta) \cos(\xi - \theta(\eta))$$

Thus the two variable expansion method to $O(\epsilon)$ yields the result that the limit cycle in van der Pol's equation is stable.

Computer Algebra

We shall present a MACSYMA program which will apply the two variable expansion method to a nonautonomous (i.e. forced) system of n coupled oscillators. We begin by giving a sample run on the van der Pol equation. Then we give a listing of the program and follow it with a discussion of the more general problem of n coupled oscillators:

```
TWOVAR();
DO YOU WANT TO ENTER NEW DATA (Y/N)
y;
NUMBER OF D.E.'S
1;
THE 1 D.E.'S WILL BE IN THE FORM:
X[I]'' + W[I]^2 X[I] = E F[I](X[1],....,X[ 1 ],T)
ENTER SYMBOL FOR X[ 1 ]
x;
ENTER W[ 1 ]
1;
ENTER F[ 1 ]
(1-x^2)*diff(x,t);
THE D.E.'S ARE ENTERED AS:
```

$$
x\,'' + x = e\,(1 - x\,)\,\frac{dx}{dt}
$$

```
THE METHOD ASSUMES A SOLUTION IN THE FORM:
X[I] = X0[I] + E X1[I]
WHERE X0[I] = A[I](ETA) COS W[I] XI + B[I](ETA) SIN W[I] XI
WHERE XI = T AND ETA = E T
REMOVAL OF SECULAR TERMS IN THE X1[I] EQS. GIVES:
```

$$
2\left(\frac{d}{d\,eta}\,(a_1)\right) + \frac{a_1\,b_1^2}{4} + \frac{a_1^3}{4} - a_1 = 0
$$

$$
-2\left(\frac{d}{d\,eta}\,(b_1)\right) - \frac{b_1^3}{4} - \frac{a_1^2\,b_1}{4} + b_1 = 0
$$

DO YOU WANT TO TRANSFORM TO POLAR COORDINATES (Y/N)
y;

$$\left[\left[\frac{d}{d\eta_1}(r_1) = \frac{r_1}{2} - \frac{r_1^3}{8}, \quad \frac{d}{d\eta_1}(\theta_1) = 0\right]\right]$$

DO YOU WANT TO SEARCH FOR RESONANT PARAMETER VALUES (Y/N)
n;

[VAX 8500 TIME = 16 SEC.]

Here is the program listing:


```
TWOVAR():=BLOCK(

CHOICE:READ("DO YOU WANT TO ENTER NEW DATA (Y/N)"),

IF CHOICE = N THEN GO(JUMP1),

/* INPUT D.E.'S */

N:READ("NUMBER OF D.E.'S"),

PRINT("THE",N,"D.E.'S WILL BE IN THE FORM:"),

PRINT("X[I]'' + W[I]^2 X[I] = E F[I](X[1],...,X[",N,"],T)"),

FOR I:1 THRU N DO

    X[I]:READ("ENTER SYMBOL FOR X[",I,"]"),

FOR I:1 THRU N DO

    DEPENDS(X[I],T),

FOR I:1 THRU N DO

    W[I]:READ("ENTER W[",I,"]"),

FOR I:1 THRU N DO

    F[I]:READ("ENTER F[",I,"]"),


JUMP1,

/* UPDATE EQS FOR SUBSTITUTION OF RESONANT VALUES ON 2ND TIME THRU */

FOR I:1 THRU N DO(

    W[I]:EV(W[I]),
```

```
    F[I]:EV(F[I])),
/* ECHO BACK THE D.E.'S */
PRINT("THE D.E.'S ARE ENTERED AS:"),
FOR I:1 THRU N DO
     PRINT(X[I],"''  +",W[I]^2*X[I],"=",E*F[I]),
PRINT("THE METHOD ASSUMES A SOLUTION IN THE FORM:"),
PRINT("X[I] = X0[I] + E X1[I]"),
PRINT("WHERE X0[I] = A[I](ETA) COS W[I] XI + B[I](ETA) SIN W[I] XI"),
PRINT("WHERE XI = T AND ETA = E T"),
/* DON'T USE A OR B AS PARAMETERS IN THE GIVEN D.E.'S */
DEPENDS([A,B],ETA),
FOR I:1 THRU N DO
     X0[I]:A[I]*COS(W[I]*XI)+B[I]*SIN(W[I]*XI),
FOR I:1 THRU N DO
     G[I]:EV(F[I],T=XI),
FOR I:1 THRU N DO(
     FOR J:1 THRU N DO
          G[I]:EV(G[I],X[J]::X0[J])),
FOR I:1 THRU N DO(
     G[I]:G[I]-2*DIFF(X0[I],XI,1,ETA,1),
     G[I]:EV(G[I],DIFF),
     G[I]:EXPAND(TRIGREDUCE(EXPAND(G[I])))),
/* COLLECT SECULAR TERMS */
FOR I:1 THRU N DO(
     S[I]:COEFF(G[I],SIN(W[I]*XI)),
     C[I]:COEFF(G[I],COS(W[I]*XI))),
/* DISPLAY SECULAR TERMS */
PRINT("REMOVAL OF SECULAR TERMS IN THE X1[I] EQS. GIVES:"),
FOR I:1 THRU N DO(
```

```
    PRINT(S[I],"= 0"),

    PRINT(C[I],"= 0")),

ABEQS:APPEND(MAKELIST(S[I],I,1,N),MAKELIST(C[I],I,1,N)),

CHOICE2:READ("DO YOU WANT TO TRANSFORM TO POLAR COORDINATES (Y/N)"),

IF CHOICE2=N THEN GO(JUMP2),

/* TRANSFORM TO POLAR COORDINATES */

DEPENDS([R,THETA],ETA),

TRANS:MAKELIST([A[I]=R[I]*COS(THETA[I]),B[I]=R[I]*SIN(THETA[I])],I,1,N),

INTEQS:EV(ABEQS,TRANS,DIFF),

RTHEQS:SOLVE(INTEQS,APPEND(MAKELIST(DIFF(R[I],ETA),I,1,N),

                 MAKELIST(DIFF(THETA[I],ETA),I,1,N))),

RTHEQS:TRIGSIMP(RTHEQS),

RTHEQS:EXPAND(TRIGREDUCE(RTHEQS)),

PRINT(RTHEQS),


JUMP2,

CHOICE3:READ("DO YOU WANT TO SEARCH FOR RESONANT PARAMETER VALUES (Y/N)"),

IF CHOICE3=N THEN GO(END),

/* OBTAIN FREQUENCIES WHICH APPEAR ON RHS'S OF D.E.'S */

/* DEFINE PATTERN MATCHING RULES TO ISOLATE FREQS */

MATCHDECLARE([DUMMY1,DUMMY2],TRUE),

DEFRULE(COSARG,DUMMY1*COS(DUMMY2),DUMMY2),

DEFRULE(SINARG,DUMMY1*SIN(DUMMY2),DUMMY2),

FOR I:1 THRU N DO(

/* CHANGE SUM TO A LIST */

    TEMP1:ARGS(G[I]),

/* REMOVE CONSTANT TERMS */

    TEMP2:MAP(TRIGIDENTIFY,TEMP1),
```

```
/* ISOLATE ARGUMENTS OF TRIG TERMS */

    TEMP3:APPLY1(TEMP2,COSARG,SINARG),

    TEMP4:EV(TEMP3,XI=1),

/* REMOVE DUPLICATE ARGUMENTS */

    FREQ[I]:SETIFY(TEMP4)),

/* DISPLAY FREQUENCIES */

FOR I:1 THRU N DO(

    PRINT(X[I],"EQ'S RESONANT FREQ =",W[I]),

    PRINT("FREQS ON RHS =",FREQ[I])),


JUMP3,

PAR:READ("WHICH PARAMETER TO SEARCH FOR ?"),

/* COMPUTE RESONANT VALUES OF DESIRED PARAMETER */

RESVALS:[],

FOR I:1 THRU N DO(

    FOR J:1 THRU LENGTH(FREQ[I]) DO(

        RES:SOLVE(PART(FREQ[I],J)=W[I],PAR),

        IF (RES#[] AND RES#ALL) THEN RESVALS:APPEND(RESVALS,RES),

        RES:SOLVE(PART(FREQ[I],J)=-W[I],PAR),

        IF (RES#[] AND RES#ALL) THEN RESVALS:APPEND(RESVALS,RES))),

/* ELIMINATE DUPLICATE PARAMETER VALUES */

RESVALUES:SETIFY(RESVALS),

/* DISPLAY RESONANT PARAMETER VALUES */

PRINT(RESVALUES),

CHOICE4:READ("DO YOU WANT TO SEARCH FOR ANOTHER PARAMETER (Y/N) ?"),

IF CHOICE4=Y THEN GO(JUMP3),

END," ")$
```

```
/* AUXILIARY FUNCTIONS */
TRIGIDENTIFY(EXP):=IF FREEOF(SIN,EXP) AND FREEOF(COS,EXP) THEN 0 ELSE EXP$


SETIFY(LIST):=(
    SET:[LIST[1]],
    FOR I:2 THRU LENGTH(LIST) DO(
        IF NOT MEMBER(LIST[I],SET) THEN SET:CONS(LIST[I],SET)),
    SET)$
```

Systems of Coupled Oscillators

Our main interest in the two variable method will be in applications to systems of n oscillators:

$$(16) \qquad \frac{d^2 x_i}{dt^2} + \omega_i^2 x_i = \epsilon f_i(x_1, x_2, \ldots, x_n, t) \ , \quad i = 1, 2, \ldots, n$$

After replacing time t by the two independent variables ξ and η,

$$(17) \qquad \xi = t, \quad \eta = \epsilon t$$

and expanding each of the dependent variables x_i in a power series in ϵ,

$$(18) \qquad x_i(\xi, \eta) = x_{i0}(\xi, \eta) + \epsilon x_{i1}(\xi, \eta) + \cdots \ , \quad i = 1, 2, \ldots, n$$

we obtain the following differential equation on x_{i1}:

$$(19) \qquad \frac{\partial^2 x_{i1}}{\partial \xi^2} + \omega_i^2 x_{i1} = -2 \frac{\partial^2 x_{i0}}{\partial \xi \, \partial \eta} + f_i(x_{10}, x_{20}, \ldots, x_{n0}, \xi)$$

where (cf. eq.(9))

(20) $x_{i0} = A_i(\eta) \cos \omega_i \xi + B_i(\eta) \sin \omega_i \xi$

     The program removes secular terms from (19) and offers the user the option of transforming to polar coordinates as in the preceding example. In addition, the program offers the user the option of searching for resonant parameter values, i.e., parameter values which cause an innocuous term on the right hand side of (19) to become a resonant term. Since we are only working to $O(\epsilon)$ throughout this chapter, the utility of this approach will be most significant in problems involving several oscillators and/or several parameters. This is in contrast to situations in which one is concerned with obtaining many terms of a series solution, in which case computer size and speed limitations usually prohibit problems which involve more than one or two parameters.

     As an example, we will consider the following system of two oscillators:

(21.1) $\dfrac{d^2 x}{dt^2} + (1 + \epsilon \, \Delta_x) \, x + \epsilon \, \upsilon \, x^3 = \epsilon \, k \, y$

(21.2) $\dfrac{d^2 y}{dt^2} + (1 + \epsilon \, \Delta_y) \, y + \epsilon \, \mu \, y \cos \omega t = 0$

     This system may be described as a Duffing oscillator (x) driven by the output from a Mathieu equation (y). The cos ωt term drives the y equation which in turn drives the x equation. The parameters are:

Δ_x , Δ_y = detuning coefficients

υ = nonlinearity coefficient

μ = forcing coefficient

ω = forcing frequency

k = coupling coefficient

    Here is the MACSYMA run annotated with comments in *italics*:

TWOVAR();

DO YOU WANT TO ENTER NEW DATA (Y/N)
y;
NUMBER OF D.E.'S
2;
THE 2 D.E.'S WILL BE IN THE FORM:
X[I]'' + W[I]^2 X[I] = E F[I](X[1],....,X[2],T)
ENTER SYMBOL FOR X[1]
x;
ENTER SYMBOL FOR X[2]
y;
ENTER W[1]
1;
ENTER W[2]
1;
ENTER F[1]
-delx*x-nu*x^3+k*y;
ENTER F[2]
-dely*y-mu*y*cos(w*t);
THE D.E.'S ARE ENTERED AS:

$$x'' + x = e\ (k\ y - nu\ x^3 - delx\ x)$$

$$y'' + y = e\ (-\ mu\ \cos(t\ w)\ y - dely\ y)$$

THE METHOD ASSUMES A SOLUTION IN THE FORM:
X[I] = X0[I] + E X1[I]
WHERE X0[I] = A[I](ETA) COS W[I] XI + B[I](ETA) SIN W[I] XI
WHERE XI = T AND ETA = E T
REMOVAL OF SECULAR TERMS IN THE X1[I] EQS. GIVES:

$$-\ \frac{3\ b_1^3\ nu}{4}\ -\ \frac{3\ a_1^2\ b_1\ nu}{4}\ +\ b_2\ k\ -\ b_1\ delx\ +\ 2\ \left(\frac{d}{deta}\ (a_1)\right)\ =\ 0$$

$$- \frac{3 \, a_1 \, b_1 \, nu}{4} - \frac{3 \, a_1^2 \, nu}{4} + a_1 \, k - a_1 \, delx - 2 \left(\frac{d}{deta} (b_1) \right) = 0$$

$$2 \left(\frac{d}{deta} (a_2) \right) - b_2 \, dely = 0$$

$$- a_2 \, dely - 2 \left(\frac{d}{deta} (b_2) \right) = 0$$

DO YOU WANT TO TRANSFORM TO POLAR COORDINATES (Y/N)
n;

We wait to transform to polars until after we have searched for resonant parameter values:

DO YOU WANT TO SEARCH FOR RESONANT PARAMETER VALUES (Y/N)
y;
x EQ'S RESONANT FREQ = 1
FREQS ON RHS = [1, 3]
y EQ'S RESONANT FREQ = 1
FREQS ON RHS = [1, w - 1, w + 1]

In order for a term on the right hand side (RHS) of eq.(19) to be resonant, its frequency must equal ω_i. The program finds all values of the selected parameter which produce a match between a resonant frequency ω_i and an element in the associated list of frequencies appearing on the RHS:

WHICH PARAMETER TO SEARCH FOR ?
w;
[w = -2, w = 0, w = 2]
DO YOU WANT TO SEARCH FOR ANOTHER PARAMETER (Y/N) ?
n;

[VAX 8500 TIME = 14 SEC.]

Having obtained a list of resonant driver frequencies w, we may investigate the effect of choosing one of them:

w:2$

We selected the resonant value 2 for w and assigned w this value. Now we again call the program TWOVAR which automatically inserts this new value of w into the previous equations:

TWOVAR();

DO YOU WANT TO ENTER NEW DATA (Y/N)
n;
THE D.E.'S ARE ENTERED AS:

$$x'' + x = e\,(k\,y - nu\,x^3 - delx\,x)$$

$$y'' + y = e\,(-\,mu\,\cos(2\,t)\,y - dely\,y)$$

THE METHOD ASSUMES A SOLUTION IN THE FORM:
X[I] = XO[I] + E X1[I]
WHERE XO[I] = A[I](ETA) COS W[I] XI + B[I](ETA) SIN W[I] XI
WHERE XI = T AND ETA = E T
REMOVAL OF SECULAR TERMS IN THE X1[I] EQS. GIVES:

$$-\frac{3\,b_1\,nu}{4} - \frac{3\,a_1^2\,b_1\,nu}{4} + b_2\,k - b_1\,delx + 2\left(\frac{d}{deta}(a_1)\right) = 0$$

$$-\frac{3\,a_1\,b_1^2\,nu}{4} - \frac{3\,a_1^3\,nu}{4} + a_2\,k - a_1\,delx - 2\left(\frac{d}{deta}(b_1)\right) = 0$$

$$\frac{b_2\,mu}{2} - b_2\,dely + 2\left(\frac{d}{deta}(a_2)\right) = 0$$

$$-\frac{a_2\,mu}{2} - a_2\,dely - 2\left(\frac{d}{deta}(b_2)\right) = 0$$

DO YOU WANT TO TRANSFORM TO POLAR COORDINATES (Y/N)
y;

$$\left[\left[\frac{d}{deta}(r_1) = -\frac{r_2\,\sin(theta_2 - theta_1)\,k}{2}, \quad \frac{d}{deta}(r_2) = -\frac{r_2\,\sin(2\,theta_2)\,mu}{4},\right.\right.$$

```
                2
             3 r  nu    r  cos(theta  - theta ) k
  d           1       2        2       1         delx
---- (theta ) = - -------- + -------------------------- - ----,
deta       1        8                  2 r                 2
                                          1


               cos(2 theta ) mu
  d                        2        dely
---- (theta ) = - ---------------- - ----]]
deta       2             4            2
```

DO YOU WANT TO SEARCH FOR RESONANT PARAMETER VALUES (Y/N)
n;

[VAX 8500 TIME = 91 SEC.]

The program has obtained the slow flow equations corresponding to the

resonant value $\omega = 2$. As always in this method there is still the problem of

dealing with the slow flow equations. Since the y equation drives the x

equation (but not vice versa), the slow flow equations on A_2 and B_2 are

uncoupled from the rest:

$$(22.1) \qquad A_2' = -\frac{1}{2}\left[\frac{\mu}{2} - \Delta_y\right] B_2$$

$$(22.2) \qquad B_2' = -\frac{1}{2}\left[\frac{\mu}{2} + \Delta_y\right] A_2$$

Differentiating (22.1) and substituting in (22.2) produces the
equation:

$$(23) \qquad A_2'' + \frac{1}{4}\left[\Delta_y^2 - \frac{\mu^2}{4}\right] A_2 = 0$$

Thus the system (22) has bounded solutions if and only if $\Delta_y^2 - \frac{\mu^2}{4} > 0$, i.e.,
if and only if

$$(24) \qquad \left|\Delta_y\right| > \left|\frac{\mu}{2}\right|$$

This yields the well known result [44] that for small forcing amplitudes, Mathieu's equation (21.2) possesses a resonance instability when the forcing frequency is approximately twice the natural frequency. In fact, an infinite number of such regions of instability exist when $\omega = 2/n$, for $n = 1, 2, 3, \ldots$, but the perturbation method must be extended to $O(\epsilon^n)$ in order to pick up the nth such region [35].

When eq.(24) holds, or when the forcing frequency ω is not resonant, the general solution (20) to Mathieu's equation (21.2) is composed of terms which are the product of a slowly varying periodic function (A_2 or B_2) and a periodic function of frequency unity (cos t or sin t). This results, in general, in an almost periodic function.

We now consider the full slow flow system of 4 differential equations in the resonant case ($\omega = 2$). In polar form these were found to be:

$$(25.1) \qquad r_1' = - \frac{r_2 k}{2} \sin(\theta_2 - \theta_1)$$

$$(25.2) \qquad \theta_1' = - \frac{3}{8} \upsilon \, r_1^2 + \frac{r_2 k}{2 r_1} \cos(\theta_2 - \theta_1) - \frac{\Delta_x}{2}$$

$$(25.3) \qquad r_2' = - \frac{r_2 \mu}{4} \sin 2\theta_2$$

$$(25.4) \qquad \theta_2' = - \frac{\mu}{4} \cos 2\theta_2 - \frac{\Delta_y}{2}$$

These equations are too complicated to hope for an exact general solution. Instead we will look for equilibria. Note that an equilibrium of the slow flow system (25) corresponds to a periodic solution of the original system (21) (cf. eq.(20)).

In order for eq.(25.3) to have an equilibrium, either $r_2 = 0$ (in which case the y oscillator is turned off so that the problem becomes the free undamped Duffing equation, which is integrable), or $\sin 2\theta_2 = 0$. In the latter case, $\cos 2\theta_2 = \pm 1$, and for equilibrium, eq.(25.4) requires $\Delta_y = \pm \frac{\mu}{2}$, a condition corresponding to the transition between stability and instability in the Mathieu equation on y (cf. eq.(24)). Requiring this condition to hold, we set the right hand sides of eqs.(25.1) and (25.2) to zero for equilibrium. Eq.(25.1) requires either $r_2 = 0$ (the integrable case again) or $\sin(\theta_2 - \theta_1) = 0$. In the latter case, $\cos(\theta_2 - \theta_1) = \pm 1$, and eq.(25.2) requires

(26) $$-\frac{3}{8} \nu r_1^2 \pm \frac{r_2 k}{2 r_1} - \frac{\Delta_x}{2} = 0$$

This equation relates the amplitude of the derived periodic motion r_1 to the amplitude of the driver r_2 and the other parameters in the case that $\Delta_y = \pm \frac{\mu}{2}$.

Exercises

1. Find the slow flow associated with the dynamics of two coupled van der Pol oscillators [37],[10]:

$$\frac{d^2 x}{dt^2} + x - \epsilon (1 - x^2) \frac{dx}{dt} = \epsilon k_{11} x + \epsilon k_{12} y + \epsilon c_{11} \frac{dx}{dt} + \epsilon c_{12} \frac{dy}{dt}$$

$$\frac{d^2 y}{dt^2} + y - \epsilon (1 - y^2) \frac{dy}{dt} = \epsilon k_{21} x + \epsilon k_{22} y + \epsilon c_{21} \frac{dx}{dt} + \epsilon c_{22} \frac{dy}{dt}$$

2. Find the slow flow associated with the dynamics of n coupled van der Pol oscillators:

$$\frac{d^2x_i}{dt^2} + x_i - \epsilon (1 - x_i^2) \frac{dx_i}{dt} = \epsilon k_{i-1} (x_{i-1} - x_i) + \epsilon k_i (x_{i+1} - x_i)$$

where $i = 1,2,\ldots,n$ and where $k_0 = k_n = 0$. The procedure here is to separately consider the cases $n = 2,3,4$, etc., until a general pattern emerges regarding the structure of the slow flow equations.

3. Apply the two variable expansion method to the equation

(P1) $$\frac{d^2x}{dt^2} + \left[\omega^2 + \epsilon \Lambda + \epsilon \cos t\right] x + \epsilon c \frac{dx}{dt} + \epsilon \alpha x^3 = 0$$

in which ω = natural frequency

Λ = detuning coefficient

c = damping coefficient

α = nonlinearity coefficient

Show that $\omega = \frac{1}{2}$ is a resonant parameter value. Obtain the associated slow flow. Find a relation between Λ and c such that the trivial solution of the slow flow (which corresponds to the rest solution $x \equiv 0$ of (P1)) is <u>unstable</u>. The nontrivial equilibria of the slow flow correspond to 2:1 subharmonic periodic solutions of (P1). Find an expression for the amplitude r of such motions as a function of Λ, c and α. For given α, find a relation between Λ and c such that 2:1 subharmonics exist.

CHAPTER 5

AVERAGING

Introduction

Like normal form transformations, averaging uses a near identity
coordinate transformation to simplify a given system of ordinary differential
equations. In contrast to our treatment of normal forms, where we applied the
method to strictly autonomous systems, we shall apply averaging to
nonautonomous systems. The coordinate transformations will be chosen so as to
transform the nonautonomous system into an autonomous one called the averaged
system.

Consider the system

(1) $\qquad \dot{x} = \epsilon\, f(x,t,\epsilon)\;, \qquad\qquad x \in \mathbb{R}^n, \quad \epsilon \ll 1$

where f is T-periodic in t. Such a system will be said to be in general form
with respect to averaging. Since $\epsilon \ll 1$, x usually evolves on a much slower
time scale than does f, which we can consider as T-periodic forcing. Therefore
it seems natural that the main influence for the time evolution of x comes from
the mean of f over one period T. This is the basic idea of averaging which
leads to the following algorithm [14],[41],[16]: Set

(2) $\qquad x = y + \epsilon\, w_1(y,t)$

and insert into (1). Then we find

$$(3) \qquad \dot{y} = [I + \epsilon\, D_y w_1]^{-1} [\epsilon\, f(y + \epsilon\, w_1, t, \epsilon) - \epsilon\, \frac{\partial w_1}{\partial t}]$$

$$(4) \qquad\qquad = \epsilon \left[g_1(y, t) - \frac{\partial w_1}{\partial t} \right] + O(\epsilon^2)$$

where D_y is the Jacobian matrix operator, and where

$$(5) \qquad\qquad g_1(y, t) = f(y, t, 0).$$

We split $g_1(y, t)$ into a mean component $\bar{g}_1(y)$ and a time dependent component $\tilde{g}_1(y, t)$ with zero mean,

$$(6) \qquad\qquad g_1(y, t) = \bar{g}_1(y) + \tilde{g}_1(y, t),$$

where

$$(7) \qquad\qquad \bar{g}_1(y) = \frac{1}{T} \int_0^T f(y, t, 0)\, dt,$$

and where $\tilde{g}_1(y, t)$ is given by the difference between eqs. (5) and (7). Then we can choose $w_1(y, t)$ such that

$$(8) \qquad\qquad \frac{\partial w_1}{\partial t} = \tilde{g}_1(y, t)$$

where the arbitrary constant of integration in (8) is chosen such that w_1 has zero mean. This gives

(9) $$\dot{y} = \epsilon \, \overline{g}_1(y) + O(\epsilon^2)$$

Eq.(9) is the first order averaged system corresponding to equation
(1). There are several theorems concerning the relation between the averaged
and the general system (for details, see e.g. [41]). Since we are mainly
interested in the computational aspects of the perturbation method we shall not
state these theorems here explicitly, but we shall try to give the general idea
of their content. One can prove that a solution $y(t)$ of the averaged system
(9) follows a true solution $x(t)$ of the general system (1) for a time of order
$1/\epsilon$ if the initial values y_0 and x_0 were close to order ϵ, i.e. if
$|y_0 - x_0| = O(\epsilon)$. Also all the qualitative local behavior of the dynamics of (9)
corresponds to the same qualitative and local behavior of periodic orbits of
(1). In particular a stable (unstable) fixed point of (9) corresponds to a
stable (unstable) limit cycle of (1), and a Hopf bifurcation giving rise to an
attracting (repelling) limit cycle in (9) corresponds to a bifurcation to a
stable (unstable) invariant torus in (1), and so on. There are additional
results concerning global information on periodic orbits of (9) for which we
refer the reader to [14].

Second Order Averaging

It frequently happens that the system (9) is degenerate and does not
provide enough information about the solutions of (1). In such a case the
averaging method can be extended to include order ϵ^2 terms by replacing the
transformation (2) with

(10) $$x = y + \epsilon \, w_1(y,t) + \epsilon^2 \, w_2(y,t)$$

Substitution of (10) into (1) gives

$$(11) \qquad \dot{y} = \epsilon \left[g_1(y,t) - \frac{\partial w_1}{\partial t} \right] + \epsilon^2 \left[g_2(y,t) - \frac{\partial w_2}{\partial t} \right] + O(\epsilon^3)$$

where $g_1(y,t) = f(y,t,0)$ as in (5), and where

$$(12) \qquad g_2(y,t) = D_y w_1 \left[\frac{\partial w_1}{\partial t} - f(y,t,0) \right] + \frac{\partial f}{\partial \epsilon}(y,t,0) + D_y f(y,t,0) \, w_1$$

We split $g_2(y,t)$ into a mean component $\bar{g}_2(y)$ and a time dependent component $\tilde{g}_2(y,t)$ with zero mean,

$$(13) \qquad g_2(y,t) = \bar{g}_2(y) + \tilde{g}_2(y,t),$$

where

$$(14) \qquad \bar{g}_2(y) = \frac{1}{T} \int_0^T \left[\frac{\partial f}{\partial \epsilon}(y,t,0) + D_y \tilde{g}_1(y,t) \, w_1 \right] dt$$

and where $\tilde{g}_2(y,t)$ is given by the difference between eqs.(12) and (14). We have chosen $w_1(y,t)$ as in (8), and have used the fact that $f(y,t,0) = g_1(y,t) = \bar{g}_1(y) + \tilde{g}_1(y,t)$ and that $D_y w_1 \, \bar{g}_1(y)$ and $D_y \bar{g}_1(y) \, w_1$ have zero mean. Then we can choose $w_2(t)$ such that

$$(15) \qquad \frac{\partial w_2}{\partial t} = \tilde{g}_2(y,t)$$

where the arbitrary constant of integration in (15) is chosen such that w_2 has zero mean. This gives

$$(16) \qquad \dot{y} = \epsilon \, \bar{g}_1(y) + \epsilon^2 \, \bar{g}_2(y) + O(\epsilon^3)$$

Although we shall only work to order ϵ^2 in this Chapter, the method of averaging can be extended in a similar fashion to include terms of order ϵ^m. See Exercise 1 at the end of this Chapter.

Van der Pol Transformation

Averaging is frequently used to treat weakly nonlinear forced oscillators of the form

$$(17) \qquad \ddot{z} + \omega_0^2 \, z = \epsilon \, F(z, \dot{z}, t, \omega, \epsilon)$$

where ω is the forcing frequency. Although this important class of applications is not of the form (1), it may be put in general form with respect to averaging by the following procedure.

The unperturbed system ($\epsilon = 0$) has the general solution

$$(18) \qquad \begin{bmatrix} z \\ \dot{z} \end{bmatrix} = \begin{bmatrix} \cos(\omega_0 t) \\ -\omega_0 \sin(\omega_0 t) \end{bmatrix} x_1 - \begin{bmatrix} \sin(\omega_0 t) \\ \omega_0 \cos(\omega_0 t) \end{bmatrix} x_2$$

where x_1 and x_2 are constants given by the initial values of the unperturbed problem. Allowing for x_1 and x_2 to be time dependent (the method of variation of parameters), and replacing ω_0 by an as yet unspecified frequency ω_1, we obtain the van der Pol transformation [14]

$$(19) \qquad \begin{bmatrix} z \\ \dot{z} \end{bmatrix} = A \begin{bmatrix} x_1 \\ x_2 \end{bmatrix} = \begin{bmatrix} \cos(\omega_1 t) & -\sin(\omega_1 t) \\ -\omega_1 \sin(\omega_1 t) & -\omega_1 \cos(\omega_1 t) \end{bmatrix} \begin{bmatrix} x_1 \\ x_2 \end{bmatrix}$$

With this transformation, eq.(17) becomes

$$\dot{x}_1 = -\frac{1}{\omega_1}\left[(\omega_1^2 - \omega_0^2)\, z + \epsilon\, F(z,\dot{z},t,\omega,\epsilon)\right]\sin(\omega_1 t)$$

(20)

$$\dot{x}_2 = -\frac{1}{\omega_1}\left[(\omega_1^2 - \omega_0^2)\, z + \epsilon\, F(z,\dot{z},t,\omega,\epsilon)\right]\cos(\omega_1 t)$$

A natural choice for ω_1 is the forcing frequency ω. However, in order to include the case of subharmonic resonance of order k, i.e. the case where $\omega = k\,\omega_0$, we choose $\omega_1 = \omega/k$. Allowing for an order ϵ detuning off of the subharmonic resonance, we set $\epsilon\,\Omega = \omega^2 - k^2\omega_0^2$. With these substitutions, (20) is in general form with respect to averaging, cf. eq.(1):

$$\dot{x}_1 = -\epsilon\frac{k}{\omega}\left[\frac{\Omega}{k^2}\, z + F(z,\dot{z},t,\omega,\epsilon)\right]\sin(\omega t/k)$$

(21)

$$\dot{x}_2 = -\epsilon\frac{k}{\omega}\left[\frac{\Omega}{k^2}\, z + F(z,\dot{z},t,\omega,\epsilon)\right]\cos(\omega t/k)$$

where z and $\dot{z}$ are given by (19).

An Example

As an example, let us take van der Pol's equation,

(22) $$\ddot{z} + z = \epsilon\,(1 - z^2)\,\dot{z}$$

Performing the van der Pol transformation (19) with $\omega_1 = \omega_0 = 1$ gives, from (20),

$$\dot{x}_1 = -\epsilon\,(1 - z^2)\,\dot{z}\,\sin t$$

(23)

$$\dot{x}_2 = -\epsilon\,(1 - z^2)\,\dot{z}\,\cos t$$

where $z = x_1 \cos t - x_2 \sin t$. Now we may apply the transformation (2), with w_1 as in (8), which turns out to give the averaged equations (cf. (9)):

(24)
$$\dot{y}_1 = -\frac{\epsilon}{8} y_1 \left[y_1^2 + y_2^2 - 4 \right]$$

$$\dot{y}_2 = -\frac{\epsilon}{8} y_2 \left[y_1^2 + y_2^2 - 4 \right]$$

Eqs.(24) may be simplified by transforming to polar coordinates,

(25) $\qquad\qquad y_1 = r \cos \theta, \quad y_2 = r \sin \theta$

which gives

(26) $\qquad\qquad \dot{r} = \frac{\epsilon}{8} \left[r^3 - 4 r \right], \quad \dot{\theta} = 0$

Equilibria for eq.(26) are $r = 0$ and $r = 2$. The latter recovers the small ϵ approximation for the limit cycle in van der Pol's equation (22).

Computer Algebra

The MACSYMA program which we shall use to implement averaging takes a system of the form (1) and transforms it to the first order averaged form (9) or the second order averaged form (16).

In the case of first order averaging this requires only that $\bar{g}_1(y)$ be calculated from eq.(7). Instead of computing the integral in (7), however, we use a more efficient scheme for eliminating the periodic terms. We convert all sines and cosines to complex exponentials (a step performed by the MACSYMA function EXPONENTIALIZE), algebraically expand the result and then kill all terms which contain exponentials in t (a step performed using a pattern matching rule called KILLEXP).

In the case of second order averaging we must also compute $\bar{g}_2(y)$ from eq.(14). This involves finding w_1 from (8). Periodic terms are again removed by using pattern matching instead of integration.

In order to conveniently treat weakly nonlinear oscillators of the form (17), we include the van der Pol transformation in the program.

Before giving a sample run and the program listing, we mention two caveats to the user:

1. The user must scale time so that averaging is performed over period 2π (except in the case of a weakly nonlinear oscillator (17), where this is done automatically.)

2. Our scheme for replacing integration by elimination of exponentials only works for that class of problems in which all trigonometric terms occur in the numerator of the vector field $f(x,t,\epsilon)$ in (1). In other cases, the program may be easily changed to use integration to compute $\bar{g}_1(y)$ and $\bar{g}_2(y)$ in (7) and (14).

Here is a sample run on the van der Pol oscillator (22):

```
AVERAGE();

DO YOU WANT TO ENTER A NEW PROBLEM? (Y/N)
Y;
ARE YOU CONSIDERING A WEAKLY NONLINEAR OSCILLATOR OF THE FORM
Z'' + OMEGA0^2 Z = EPS F(Z,ZDOT,T) ? (Y/N)
Y;
ENTER OMEGA0
1;
ENTER F(Z,ZDOT,T)
(1-Z^2)*ZDOT;
DO YOU WANT FIRST OR SECOND ORDER AVERAGING?   (1/2)
1;
                  2      3                  3      2
            y1 y2      y1      y1         y2     y1 y2    y2
  [eps (- ------- - --- + --),  eps (- --- - ------ + --)]
             8        8      2           8       8       2

[VAX 8500 TIME = 25 SEC.]
```

This result is the same as eq.(24) given in the text. Before converting to polar coordinates (25), we run the program again to include second order effects. Note that the problem need not be reentered:

AVERAGE();

DO YOU WANT TO ENTER A NEW PROBLEM? (Y/N)
N;
DO YOU WANT FIRST OR SECOND ORDER AVERAGING?   (1/2)
2;

$$\text{matrix}\left(\left[\text{eps }\left(\frac{2\;11\;y2}{256} + \frac{11\;y1^5\;y2^2}{128} - \frac{3\;y2^3}{16} + \frac{11\;y1^4\;y2}{256} - \frac{3\;y1\;y2^2}{16} + \frac{y2^2}{8}\right)\right.\right.$$

$$\left.+ \text{eps }\left(- \frac{y1\;y2^2}{8} - \frac{y1^3}{8} + \frac{y1}{2}\right)\right], \left[\text{eps }\left(- \frac{11\;y1\;y2^4}{256} - \frac{11\;y1^3\;y2^2}{128} + \frac{3\;y1\;y2^2}{16}\right.\right.$$

$$\left.\left.\left.- \frac{11\;y1^5}{256} + \frac{3\;y1^3}{16} - \frac{y1}{8}\right) + \text{eps }\left(- \frac{y2^3}{8} - \frac{y1\;y2^2}{8} + \frac{y2^2}{2}\right)\right]\right)$$

[VAX 8500 TIME = 68 SEC.]

In order to transform to polar coordinates, we use the program

TRANSFORM given in Chapter 2. The output of the program AVERAGE is stored in a

variable called RESULT:

TRANSFORM();
Enter number of equations
2;
Enter symbol for original variable 1
Y1;
Enter symbol for original variable 2
Y2;
Enter symbol for transformed variable 1
R;
Enter symbol for transformed variable 2
THETA;
The RHS's of the d.e.'s are functions of the original variables:
Enter RHS of y1 d.e.
d y1 /dt =
RESULT[1,1];

$$\text{d y1 /dt} = \text{eps }\left(\frac{2\;11\;y2}{256} + \frac{11\;y1^5\;y2^2}{128} - \frac{3\;y2^3}{16} + \frac{11\;y1^4\;y2}{256} - \frac{3\;y1\;y2^2}{16} + \frac{y2^2}{8}\right)$$

$$+ \text{eps }\left(- \frac{y1\;y2^2}{8} - \frac{y1^3}{8} + \frac{y1}{2}\right)$$

Enter RHS of y2 d.e.
d y2 /dt =
RESULT[2,1];

$$d\ y2\ /dt = eps^2\ (-\frac{11\ y1\ y2^4}{256} - \frac{11\ y1^3\ y2^2}{128} + \frac{3\ y1\ y2^2}{16} - \frac{11\ y1^5}{256} + \frac{3\ y1^3}{16} - \frac{y1}{8})$$

$$+ eps\ (-\frac{y2^3}{8} - \frac{y1^2\ y2}{8} + \frac{y2}{2})$$

The transformation is entered next:
Enter y1 as a function of the new variables
y1 =
R*COS(THETA);
y1 = r cos(theta)
Enter y2 as a function of the new variables
y2 =
R*SIN(THETA);
y2 = r sin(theta)

$$[[\frac{dr}{dt} = -\frac{eps\ r^3\ sin^2(theta) + eps\ r^3\ cos^2(theta) - 4\ eps\ r}{8},$$

$$\frac{dtheta}{dt} = -(11\ eps^2\ r^4\ sin^4(theta) + (22\ eps^2\ r^4\ cos^2(theta) - 48\ eps^2\ r^2)$$

$$sin^2(theta) + 11\ eps^2\ r^4\ cos^4(theta) - 48\ eps^2\ r^2\ cos^2(theta) + 32\ eps^2)/256]]$$

[VAX 8500 TIME = 2 SEC.]

In order to simplify the algebra here we use the MACSYMA function

TRIGSIMP:

TRIGSIMP(%);

$$[[\frac{dr}{dt} = -\frac{eps\ r^3 - 4\ eps\ r}{8},\ \frac{dtheta}{dt} = -\frac{11\ eps^2\ r^4 - 48\ eps^2\ r^2 + 32\ eps^2}{256}]]$$

[VAX 8500 TIME = 13 SEC.]

This result is the $O(\epsilon^2)$ version of eq.(26). Note that on the limit

cycle r = 2, this result gives $\dot\theta = -\frac{\epsilon^2}{16}$, in agreement with the frequency

$1 - \frac{\epsilon^2}{16}$ given by Lindstedt's method. cf. Chapter 1.

Here is the program listing:

```
/* PROGRAM TO PERFORM 1ST OR 2ND ORDER AVERAGING

   ON AN N-DIMENSIONAL SYSTEM OF NONAUTONOMOUS ODE'S */

/* AVERAGING IS PERFORMED BY CONVERTING TRIG TERMS TO

   COMPLEX EXPONENTIALS, THEN KILLING EXPONENTIALS  */

AVERAGE():=BLOCK(

     CHOICE1:READ("DO YOU WANT TO ENTER A NEW PROBLEM? (Y/N)"),

     IF CHOICE1 = N THEN GO(JUMP1),

     KILL(X),

     PRINT("ARE YOU CONSIDERING A WEAKLY NONLINEAR OSCILLATOR OF THE FORM"),

     CHOICE2:READ("Z'' + OMEGAO^2 Z = EPS F(Z,ZDOT,T) ? (Y/N)"),

     IF CHOICE2 = N THEN GO(JUMP2),

/* ENTER DATA FOR SINGLE OSCILLATOR PROBLEM */

     N:2,

     OMEGAO:READ("ENTER OMEGAO"),

     F:READ("ENTER F(Z,ZDOT,T)")*EPS,

/* DOES F(Z,ZDOT,T) DEPEND EXPLICITLY ON T? */

     TEST:DIFF(F,T),

     IF TEST=0 THEN OMEGA1:OMEGAO

          ELSE(

          OMEGA:READ("ENTER THE FORCING FREQUENCY"),

          K:READ("ENTER THE ORDER OF THE SUBHARMONIC RESONANCE"),

          OMEGA1:OMEGA/K),

/* VAN DER POL TRANSFORMATION */

     ZSUB:[Z=COS(OMEGA1*T)*X1-SIN(OMEGA1*T)*X2,

               ZDOT=-OMEGA1*SIN(OMEGA1*T)*X1-OMEGA1*COS(OMEGA1*T)*X2],

/* SUBSTITUTE ZSUB INTO TRANSFORMED EQS */

/* SCALE TIME SO THAT AVERAGING OCCURS OVER PERIOD 2 PI */
```

```
    VF:EV([-1/OMEGA1^2*(EPS*KAPOMEGA/K^2*Z + F)*SIN(OMEGA1*T),

            -1/OMEGA1^2*(EPS*KAPOMEGA/K^2*Z + F)*COS(OMEGA1*T)],

            ZSUB,T=TAU/OMEGA1,INFEVAL),

    VF:EV(VF,TAU=T),

    IF OMEGA1 # OMEGAO

    THEN PRINT("WE WRITE EPS*KAPOMEGA =",OMEGA^2-K^2*OMEGAO^2)

        ELSE VF:EV(VF,KAPOMEGA=0),

    VF2:EXPAND(EXPONENTIALIZE(VF)),

    FOR I:1 THRU 2 DO VF2[I]:MAP(FACTOR,VF2[I]),

    X:[X1,X2],

    GO(JUMP1),


JUMP2,

/* ENTER DATA FOR GENERAL PROBLEM OF N ODE'S */

    N:READ("ENTER NUMBER OF DIFFERENTIAL EQUATIONS"),

    X:MAKELIST(CONCAT('X,I),I,1,N),

    PRINT("THE ODE'S ARE OF THE FORM:"),

    PRINT("DX/DT = EPS F(X,T)"),

    PRINT("WHERE X =",X),

    PRINT("SCALE TIME T SUCH THAT AVERAGING OCCURS OVER 2 PI"),

    VF:MAKELIST(READ("ENTER RHS(",I,")=EPS*...")*EPS,I,1,N),

    FOR I:1 THRU N DO PRINT("D",X[I],"/DT =",VF[I]),

    VF2:EXPAND(EXPONENTIALIZE(VF)),

    FOR I:1 THRU N DO VF2[I]:MAP(FACTOR,VF2[I]),


JUMP1,

/* AVERAGING */

    M:READ("DO YOU WANT FIRST OR SECOND ORDER AVERAGING?  (1/2)"),

    COEFVFEPS1:COEFF(VF2,EPS),
```

```
COEFVFEPS2:COEFF(VF2,EPS,2),

G1BAR:DEMOIVRE(APPLY1(COEFVFEPS1,KILLEXP)),

IF M=1 THEN RESULT:EPS*G1BAR

        ELSE(

        G1TILDE:COEFVFEPS1-G1BAR,

        W1:INTEGRATE(G1TILDE,T),

REMARRAY(JACOB),

JACOB[I,J] := DIFF(G1TILDE[I],X[J]),

JACG1TILDE:GENMATRIX(JACOB,N,N),

G2BAR:DEMOIVRE(APPLY1(EXPAND(JACG1TILDE.W1)+COEFVFEPS2,KILLEXP)),

RESULT:EPS*G1BAR+EPS^2*G2BAR),
```
/* REPLACE X BY Y */
```
FOR I:1 THRU N DO RESULT:SUBST(CONCAT('Y,I),CONCAT('X,I),RESULT),

RESULT)$
```


/* AUXILIARY FUNCTIONS TO KILL EXPONENTIAL TERMS */
```
CONTAINS_T(EXP):= NOT FREEOF(T,EXP)$

MATCHDECLARE(Q,CONTAINS_T)$

DEFRULE(KILLEXP,EXP(Q),0)$
```

Additional Examples

As our second example, we consider a modification of the van der Pol
equation (22) (from [41]):

$$(27) \qquad \ddot{z} + z = \epsilon \, \alpha \, z^2 + \epsilon^2 \, (1-z^2) \, \dot{z}$$

Here the first order mean is zero, and we must appeal to second order averaging
for a nontrivial result. Application of our program AVERAGE as in the

preceding example gives the RESULT:

```
[                 2  3            2            2   2          3         ]
[       2  5 alpha  y2       y1 y2      5 alpha  y1  y2      y1      y1  ]
[  eps  (------------- - ------- + ---------------- - --- + --)        ]
[             12           8            12            8     2          ]
[                                                                      ]
[                 3            2    2         2                  2   3  ]
[       2       y2     5 alpha  y1 y2      y1  y2      y2   5 alpha  y1 ]
[  eps  (- --- - ---------------- - ------- + -- - -------------)      ]
[          8           12             8       2         12            ]
```

and transformation to polar coordinates using the program TRANSFORM gives:

$$\left[\left[\frac{dr}{dt} = -\frac{eps^2 r^3 - 4 eps r}{8}, \quad \frac{dtheta}{dt} = -\frac{5 alpha^2 eps^2 r^2}{12}\right]\right]$$

As an example of a nonautonomous system, we take the forced damped
Duffing equation:

$$(28) \qquad\qquad \ddot{z} + z = \epsilon \left[\gamma \cos \omega t - \delta \dot{z} - \alpha z^3 \right]$$

This equation is the natural symmetric extension of the forced damped harmonic
oscillator and thus it occurs in a wide variety of applications. Here is a
sample run of our program AVERAGE applied to eq.(28):

```
AVERAGE();

DO YOU WANT TO ENTER A NEW PROBLEM? (Y/N)
Y;
ARE YOU CONSIDERING A WEAKLY NONLINEAR OSCILLATOR OF THE FORM
Z'' + OMEGAO^2 Z = EPS F(Z,ZDOT,T) ? (Y/N)
Y;
ENTER OMEGAO
1;
ENTER F(Z,ZDOT,T)
GAMMA*COS(W*T)-DELTA*ZDOT-ALPHA*Z^3;
ENTER THE FORCING FREQUENCY
W;
ENTER THE ORDER OF THE SUBHARMONIC RESONANCE
1;
```

WE WRITE EPS*KAPOMEGA = $w^2 - 1$
DO YOU WANT FIRST OR SECOND ORDER AVERAGING?   (1/2)
2;

$$\text{matrix}([eps^2 \ (- \frac{3\ alpha\ y1\ y2\ gamma}{16\ w^4} + \frac{delta\ gamma}{8\ w^3} + \frac{51\ alpha^2\ y2^5}{256\ w^4}$$

$$+ \frac{51\ alpha^2\ y1^2\ y2^3}{128\ w^4} - \frac{3\ alpha\ kapomega\ y2^3}{8\ w^4} + \frac{51\ alpha^2\ y1^4\ y2}{256\ w^4}$$

$$- \frac{3\ alpha\ kapomega\ y1^2\ y2}{8\ w^4} + \frac{delta\ y2^2}{8\ w^2} + \frac{kapomega^2\ y2}{8\ w^4})$$

$$+ eps\ (- \frac{3\ alpha\ y2^3}{8\ w^2} - \frac{3\ alpha\ y1^2\ y2}{8\ w^2} + \frac{kapomega\ y2}{2\ w^2} - \frac{delta\ y1}{2\ w})],$$

$$[eps^2\ (- \frac{3\ alpha\ y2^2\ gamma}{32\ w^4} + \frac{9\ alpha\ y1^2\ gamma}{32\ w^4} - \frac{kapomega\ gamma}{8\ w^4}$$

$$- \frac{51\ alpha^2\ y1\ y2^4}{256\ w^4} - \frac{51\ alpha^2\ y1^3\ y2^2}{128\ w^4} + \frac{3\ alpha\ kapomega\ y1\ y2^2}{8\ w^4}$$

$$- \frac{51\ alpha^2\ y1^5}{256\ w^4} + \frac{3\ alpha\ kapomega\ y1^3}{8\ w^4} - \frac{delta\ y1^2}{8\ w^2} - \frac{kapomega^2\ y1}{8\ w^4})$$

$$+ eps\ (- \frac{gamma}{2\ w^2} + \frac{3\ alpha\ y1\ y2^2}{8\ w^2} - \frac{delta\ y2}{2\ w} + \frac{3\ alpha\ y1^3}{8\ w^2} - \frac{kapomega\ y1}{2\ w^2})]])$$

[VAX 8500 TIME = 162 SEC.]

These averaged equations may be simplified by transforming to polar
coordinates (25) via the program TRANSFORM, as in the previous sample run.

Neglecting terms of $O(\epsilon^2)$ for simplicity, we obtain:

$$\left[\left[\frac{dr}{dt} = -\frac{(\text{delta } r\, w + \sin(\text{theta})\, \text{gamma})\, \text{eps}}{2\, w^2},\right.\right.$$

$$\left.\left.\frac{d\text{theta}}{dt} = \frac{(3\,\text{alpha } r^3 - 4\,\text{kapomega } r - 4\cos(\text{theta})\,\text{gamma})\,\text{eps}}{8\, r\, w}\right]\right]$$

An equilibrium of this averaged system corresponds to a limit cycle in the original system (28). Thus we set $\dot{r} = \dot{\theta} = 0$ in these last equations, solve respectively for $\sin\theta$ and $\cos\theta$, and use the identity $\sin^2\theta + \cos^2\theta = 1$ to obtain an equation on the amplitude r of the limit cycle:

$$\frac{\text{delta}^2\, r^2\, w^2}{\text{gamma}^2} + \frac{(3\,\text{alpha } r^3 - 4\,\text{kapomega } r)^2}{16\,\text{gamma}^2} = 1$$

See Fig.10.

The purpose of our next example is to offer a check on our program by comparison with previously published second order averaging computations done by hand. Holmes and Holmes [17] studied the stability and bifurcation of subharmonic periodic solutions in the equation:

(29) $$\ddot{u} - u + u^3 = \epsilon\,\gamma\cos\omega t - \epsilon\,\delta\,\dot{u}$$

Note that when $\epsilon = 0$, eq.(29) has equilibria at $u = 0, \pm1$. The equilibrium at $u = 0$ is an unstable saddle, while those at ±1 are stable centers.

Let $u = U(t)$ be a periodic solution to (29) and set

(30) $$u = U(t) + \epsilon z$$

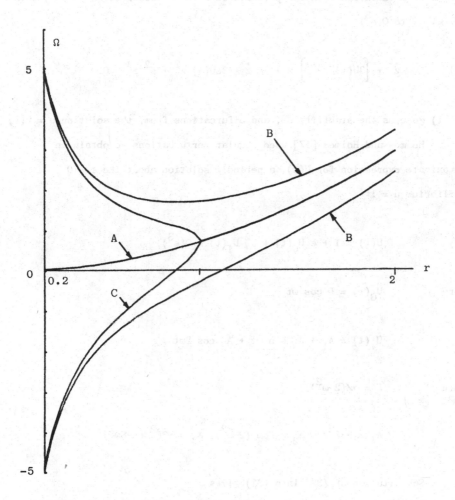

Fig. 10. Averaging $O(\epsilon)$ solution for the steady state periodic response of the forced damped Duffing eq. (28). Ω = detuning, r = response amplitude. Using $\omega^2 \approx 1 + \Omega \epsilon$, we plot

$$\Omega = \frac{3}{4} \alpha r^2 \pm \sqrt{\frac{\gamma^2}{r^2} - \delta^2}$$

for three cases: A ($\gamma = 0$, $\delta = 0$, $\alpha = 1$)

B ($\gamma = 1$, $\delta = 0$, $\alpha = 1$)

C ($\gamma = 1$, $\delta = 1$, $\alpha = 1$)

where z is a variation from the periodic solution. Substitution of (30) into (29) gives to $O(\epsilon^3)$:

$$(31) \qquad \ddot{z} + \left[3U(t)^2 - 1\right] z + \delta\epsilon \, \dot{z} + 3\epsilon U(t) \, z^2 + \epsilon^2 \, z^3 = 0$$

Eq.(31) governs the stability of, and bifurcations from, the solution $u = U(t)$.

Holmes and Holmes [17] used regular perturbations to obtain an approximate expression for $U(t)$, a periodic solution about the $\epsilon = 0$ equilibrium $u = 1$:

$$(32) \qquad U(t) = 1 + \epsilon \, U_0(t) + \epsilon^2 \, U_1(t) + O(\epsilon^3)$$

where $U_0(t) = \Gamma \cos \omega t$

$$U_1(t) = A_1 + A_2 \sin \omega t + A_3 \cos 2\omega t$$

where $\Gamma = \gamma/(2-\omega^2)$

$$A_1 = -3\Gamma^2/4, \ A_2 = \Gamma\delta\omega/(2-\omega^2), \ A_3 = -3\Gamma^2/(4-8\omega^2)$$

Substitution of (32) into (31) gives

$$(33) \qquad \ddot{z} + 2 z = \epsilon \, (- \delta \, \dot{z} - 6\Gamma \, z \cos \omega t - 3 \, z^2) +$$

$$\epsilon^2 \left[-(3\Gamma^2 \cos^2\omega t + 6 \left[-3\Gamma^2/4 + \frac{\Gamma\delta\omega}{2-\omega^2} \sin \omega t \right.\right.$$

$$\left.\left. - \frac{3\Gamma^2}{4-8\omega^2} \cos \omega t]) \, z - 3\Gamma \, z^2 \cos \omega t - z^3 \right] + O(\epsilon^3)$$

Holmes and Holmes [17] used second order averaging on eq.(33) to
investigate the bifurcation of subharmonics of order 2, i.e. the phenomenon of
period doubling. We replicate their "tedious but elementary calculations" in
the following run of the program AVERAGE. We begin by assigning the right-hand
side of eq.(33) to a variable called RHS:

```
RHS:-DEL*ZDOT-6*GAM*COS(W*T)*Z-3*Z^2+
EPS*(-(3*GAM^2*COS(W*T)^2+6*(-3*GAM^2/4+GAM*DEL*W/(2-W^2)*SIN(W*T)
-3*GAM^2/(4-8*W^2)*COS(2*W*T)))*Z-3*GAM*COS(W*T)*Z^2-Z^3);
```

$$
eps \left(- z \left(6 \left(\frac{gam\ w\ \sin(t\ w)\ del}{2 - w^2} - \frac{3\ gam^2\ \cos(2\ t\ w)}{4 - 8\ w^2} - \frac{3\ gam^2}{4} \right) \right. \right.
$$

$$
\left. \left. + 3\ gam^2\ \cos^2(t\ w) \right) - z^3 - 3\ gam\ \cos(t\ w)\ z^2 \right) - zdot\ del - 3\ z^2
$$

$$
- 6\ gam\ \cos(t\ w)\ z
$$

```
AVERAGE();

DO YOU WANT TO ENTER A NEW PROBLEM? (Y/N)
Y;
ARE YOU CONSIDERING A WEAKLY NONLINEAR OSCILLATOR OF THE FORM
Z'' + OMEGAO^2 Z = EPS F(Z,ZDOT,T) ? (Y/N)
Y;
ENTER OMEGAO
SQRT(2);
ENTER F(Z,ZDOT,T)
RHS;
ENTER THE FORCING FREQUENCY
W;
ENTER THE ORDER OF THE SUBHARMONIC RESONANCE
2;
```

$$
\text{WE WRITE EPS*KAPOMEGA} = w^2 - 8
$$

```
DO YOU WANT FIRST OR SECOND ORDER AVERAGING?   (1/2)
2;
```

$$
matrix([eps \left(\frac{2\ y2\ del}{2\ w^2} - \frac{6\ gam\ y1\ del}{w\ (w^2 - 2)} - \frac{6\ gam\ y1\ del}{w^3} - \frac{3\ y2^3}{2\ w^2} + \frac{60\ y2^3}{w^4} \right.
$$

$$
\left. - \frac{3\ y1^2\ y2}{2\ w^2} + \frac{60\ y1^2\ y2}{w^4} + \frac{6\ gam\ y2^2}{w^2} + \frac{kapomega\ y2}{8\ w^4} + \frac{3\ gam\ kapomega\ y2}{w^4} \right.
$$

$$+ \frac{9 \text{ gam } y2^2}{w^4}) + eps\ (- \frac{y1 \text{ del}}{w} + \frac{\text{kapomega } y2}{2\,w^2} + \frac{6 \text{ gam } y2}{w^2})],$$

$$[eps^2\ (- \frac{y1 \text{ del}}{2\,w^2} + \frac{6 \text{ gam } y2 \text{ del}}{w^2\,(w^2 - 2)} + \frac{6 \text{ gam } y2 \text{ del}}{w^3} + \frac{3 \text{ y1 } y2^2}{2\,w^2} - \frac{60 \text{ y1 } y2^2}{w^4} + \frac{3 \text{ y1}^3}{2\,w^2}$$

$$- \frac{60 \text{ y1}^3}{w^4} - \frac{6 \text{ gam}^2 \text{ y1}}{w^2} - \frac{\text{kapomega}^2 \text{ y1}}{8\,w^4} + \frac{3 \text{ gam kapomega y1}}{w^4} - \frac{9 \text{ gam}^2 \text{ y1}}{w^4})$$

$$+ eps\ (- \frac{y2 \text{ del}}{w} - \frac{\text{kapomega y1}}{2\,w^2} + \frac{6 \text{ gam y1}}{w^2})])$$

[VAX 8500 TIME = 271 SEC.]

These averaged equations agree with eq.(25) in [17] (but note that our
definitions for t and Ω are different from those in [17].)

   The equilibria of the averaged equations correspond to periodic motions
for the unaveraged system (33). The averaged equations always possess an
equilibrium at the origin, corresponding to the solution u = U(t). Any
additional equilibria correspond to subharmonics of order 2. Note that the
order ϵ terms in the averaged equations are linear in y_1 and y_2, and thus
cannot account for such additional equilibria. Hence in order to investigate
period doubling, it is necessary to go to $O(\epsilon^2)$ in this example. We refer the
reader to [17] for further analysis of this system.

Exercises

1. One may envision bifurcation problems where second order averaging still leads to a degenerate autonomous dynamical system. In such a case a program to perform m-th order averaging would be welcome. It is not too difficult to write such a program bearing in mind that in order to find the m-th order mean $\bar{g}_m(y)$ we have to find the transformation

$$x = y + \epsilon \, w_1 + \epsilon^2 w_2 + \ldots + \epsilon^{m-1} w_{m-1}$$

Sketch of a solution:

One way to organize such a procedure is as follows: Since we are interested in the m-th order mean we can work with the Taylor expansion of eq.(3) up to order m:

$$[I + \epsilon \, D_y w_1]^{-1} \left[\epsilon \, f(y + \epsilon \, w_1, t, \epsilon) - \epsilon \, \frac{\partial w_1}{\partial t} \right]$$

$$(P1) \qquad \approx \left[\sum_{i=0}^{m-1} (-\epsilon)^i \, (D_y \, w_1)^i \right] \left[\epsilon \, f(y + \epsilon \, w_1, t, \epsilon) - \epsilon \, \frac{\partial w_1}{\partial t} \right]$$

Performing first order averaging on (P1) gives $\bar{g}_1$ and w_1. Substituting the latter back into (P1) gives a new vector field $f(y, t, \epsilon)$ which we now have to average to second order. This time we determine the Taylor expansion to order m of

$$[I + \epsilon^2 D_y w_2]^{-1} \left[\epsilon \, f(y + \epsilon^2 w_2, t, \epsilon) - \epsilon^2 \, \frac{\partial w_2}{\partial t} \right]$$

and so on. A listing of the crucial portion of the function AVERAGE() is given below [2]:

JUMP1,

/* AVERAGING */

      M:READ("ENTER ORDER OF AVERAGING"),

      IF M=1

/* FIRST ORDER AVERAGING */

      THEN (TEMPO:DEMOIVRE(APPLY1(VF2,KILLEXP)),

          RESULT:TAYLOR(TEMPO,EPS,0,1))

/* AVERAGING OF ORDER M > 1 */

      ELSE

      Y:MAKELIST(CONCAT('Y,I),I,1,N),

      WLIST:MAKELIST(CONCAT('W,I),I,1,N),

      DEPENDS(WLIST,CONS(T,Y)),

      TRAFO:Y,

      XSUB:MAPLIST("=",Y,X),

/* WNULL SETS WLIST TO ZERO */

      WNULL:VFUN(WLIST,0),

      JACOB[I,J] := DIFF(WLIST[I],Y[J]),

      JAC:GENMATRIX(JACOB,N,N),

/* MAIN LOOP */

      FOR I :1 THRU M-1 DO (

          TEMP1:MAPLIST("=",X,Y+EPS^I*WLIST),

/* HERE X IS THE LIST [X1,X2,...,XN], Y IS THE LIST [Y1,Y2,...,YN],

      WLIST IS THE TRANSFORMATION VECTOR [W1,W2,...,WN] */

          TEMP2:TAYLOR(SUM((-EPS)^(I*J)*JAC^^J,J,0,M-1),

                         (EV(VF2,TEMP1) - DIFF(WLIST,T)*EPS^I),EPS,0,M),

/* JAC IS THE JACOBIAN D WLIST/DY OF THE TRANSFORMATION WLIST */

          TEMP3:MATTOLIST(TEMP2,N),

          TEMP4:MAP(EXPAND,TAYLOR(EV(TEMP3,WNULL,DIFF),EPS,0,I)),

/* THE ITH ORDER MEAN */

```
        MEAN:APPLY1(TEMP4,KILLEXP),

        TEMP6:EXPAND(TEMP4-MEAN),

        TEMP7:EV(INTEGRATE(TEMP6,T),EPS=1),
/* THE ITH ORDER TRANSFORMATION */
        TEMP8:MAPLIST("=",WLIST,TEMP7),

        TEMP9:RATSIMP(TEMP8),
/* THE TRANSFORMED DE */
        VF2:EXPAND(EV(TEMP3,TEMP9,DIFF,XSUB,INFEVAL)),
/* THE SUM OF ALL TRANSFORMATIONS */
        TRAFO:EXPAND(TRAFO+EV(EPS^I*WLIST,TEMP9))),
/* END OF MAIN LOOP */
    PRINT("THE TRANSFORMATION: ",X,"="),

    PRINT(RATSIMP(DEMOIVRE(TRAFO))),
/* THE FINAL AVERAGING */
    RESULT:APPLY1(VF2,KILLEXP),
/* REPLACE X BY Y */
    FOR I:1 THRU N DO RESULT:SUBST(CONCAT('Y,I),CONCAT('X,I),RESULT),
    RESULT)$


/* AUXILIARY FUNCTIONS */
    VFUN(LIST,VALUE):=MAP(LAMBDA([U],U=VALUE),LIST)$
    MATTOLIST(MAT,DIM):=IF DIM>1 THEN MAKELIST(MAT[I,1],I,1,DIM) ELSE [MAT]$
```

For most practical purposes this program is probably too slow and
creates intermediate expressions which are too large. However it works for a
one-dimensional problem like the following:

AVERAGE( );

DO YOU WANT TO ENTER A NEW PROBLEM? (Y/N)
Y;
ARE YOU CONSIDERING A WEAKLY NONLINEAR OSCILLATOR OF THE FORM
Z'' + OMEGAO^2 Z = EPS F(Z,ZDOT,T) ? (Y/N)
N;
ENTER NUMBER OF DIFFERENTIAL EQUATIONS
1;
THE ODE'S ARE OF THE FORM:
DX/DT = EPS F(X,T)
WHERE X = [x1]
SCALE TIME T SUCH THAT AVERAGING OCCURS OVER 2 PI
ENTER RHS(1)=EPS*...
(X1-X1^2)*SIN(T)^2;

$$D\ x1\ /DT = eps\ \sin^2(t)\ (x1 - x1^2)$$

ENTER ORDER OF AVERAGING
3;
THE TRANSFORMATION: [x1] =

$$[- ((2\ eps^2\ \cos(4\ t) - 8\ eps^2\ \cos(2\ t))\ y1^3$$

$$+ (- 3\ eps^2\ \cos(4\ t) - 16\ eps\ \sin(2\ t) + 12\ eps^2\ \cos(2\ t))\ y1^2$$

$$+ (eps^2\ \cos(4\ t) + 16\ eps\ \sin(2\ t) - 4\ eps^2\ \cos(2\ t) - 64)\ y1)/64]$$

$$[- \frac{eps^3\ y1^4}{64} + \frac{eps^3\ y1^3}{32} - \frac{eps^3\ y1^3}{64} - \frac{eps^2\ y1^2}{2} + \frac{eps\ y1}{2}]$$

[VAX 8500 TIME = 48 SEC.]

2. Find a solution of the undamped Duffing equation ((28) with $\delta = 0$) via Lindstedt's method up to first order in ϵ. Show that to this order the amplitude of the periodic solution is identical to the steady state solution of the first order averaged system of (28) with $\delta = 0$.

Hint: We showed that the latter is of the form:

$$\frac{3 \alpha r^3 - 4 \Omega r}{4 \gamma} = 1$$

where r is the amplitude of the limit cycle and $\Omega = (\omega^2 - 1)/\epsilon$. Use Lindstedt's method to obtain $y_0(\omega t) = A \cos \omega t$, where $\omega = 1 + k_1 \epsilon$, and where

$$k_1 = - \frac{\gamma}{2 A} + \frac{3}{8} A^2$$

Introduction

The method of Lie transforms is a normal form method for Hamiltonian systems. Like the method of normal forms (Chapter 3), Lie transforms involves finding a change of variables so that the system of differential equations becomes simpler. That is, the perturbation expansions are performed on the transformation of coordinates rather than on the solution as a function of time (as, e.g., in Lindstedt's method.)

The difference between Lie transforms and normal forms lies in the special nature of Hamiltonian systems. Such systems occur extensively in physics, engineering, astronomy and classical mechanics, and normally involve models which <u>do not dissipate energy</u> through damping. The differential equations for a Hamiltonian system are generated by a single scalar quantity, the Hamiltonian, which often can be identified with the total energy of the system. The equations take the form:

(1.1)
$$\frac{dq_i}{dt} = \frac{\partial H}{\partial p_i}$$

(1.2)
$$\frac{dp_i}{dt} = -\frac{\partial H}{\partial q_i}$$

where $H = H(q_i, p_i, t)$ is the Hamiltonian, and where $q_i(t)$ and $p_i(t)$ are the dependent variables describing the state of the system. For a system with n degrees of freedom, i goes from 1 to n.

As an example of a Hamiltonian system, let us take a simple harmonic oscillator. In this case the Hamiltonian is:

$$(2) \qquad H(q_1, p_1) = \frac{1}{2} \left[p_1^2 + \omega^2 q_1^2 \right]$$

whereupon the differential equations of motion become:

$$(3.1) \qquad \frac{dq_1}{dt} = p_1$$

$$(3.2) \qquad \frac{dp_1}{dt} = -\omega^2 q_1$$

where q_1 corresponds to the displacement of the oscillator, and p_1 to its velocity. (The q_i and p_i are traditionally called the generalized coordinates and momenta, respectively.)

In the method of Lie transforms, we are concerned with generating a change of coordinates which will simplify the Hamiltonian, whereas in the method of normal forms we deal directly with the differential equations of motion. All quantities, including the original Hamiltonian H and the change of coordinates, are expanded in a power series in a small parameter ϵ:

$$(4) \qquad H = H_0 + \epsilon H_1 + \epsilon^2 H_2 + \cdots$$

The method generates a _canonical_ coordinate transformation , i.e., it preserves the Hamiltonian form of the equations. Thus the original system based on H with variables q_i, p_i is transformed into a new system with a new Hamiltonian K with variables Q_i, P_i. The transformation itself is specified by a _generating_ function W. Both K and W are expanded in perturbation series:

(5) $K = K_0 + \epsilon K_1 + \epsilon^2 K_2 + \cdots$

(6) $W = W_1 + \epsilon W_2 + \epsilon^2 W_3 + \cdots$

Up to terms of order ϵ^n, the method gives that K_n depends upon W_n and other

quantities which are already known. The strategy of the method is to choose W_n

so as to simplify K_n as much as possible. As usual in normal form methods, the

choice of the final canonical form K is strongly dependent on the specific

details of the problem at hand.

We shall now present the explicit details of the method. We refer the

reader to [7],[24] for a derivation of the following results. We consider a

system with N degrees of freedom. Let us denote the Poisson bracket of two

functions $f(Q_i, P_i)$ and $g(Q_i, P_i)$ by $\{f,g\}$:

(7) $$\{f,g\} = \sum_{i=1}^{N} \frac{\partial f}{\partial Q_i} \frac{\partial g}{\partial P_i} - \frac{\partial f}{\partial P_i} \frac{\partial g}{\partial Q_i}$$

We define the operators L_n and S_n as follows:

(8) $L_n = \{W_n, \ \}$

(9.1) $S_0 = \text{Id}$ (the identity operator)

(9.2) $S_n = \frac{1}{n} \sum_{m=0}^{n-1} L_{n-m} S_m$, $n = 1,2,3,\ldots$

Then the near-identity change of variables from (q_i, p_i) to (Q_i, P_i) is given by

$$(10.1) \qquad q_i = \left[S_0 + \epsilon \, S_1 + \epsilon^2 \, S_2 + \cdots \right] Q_i$$

$$(10.2) \qquad p_i = \left[S_0 + \epsilon \, S_1 + \epsilon^2 \, S_2 + \cdots \right] P_i$$

and the nth term K_n of the resulting Hamiltonian K of eq.(5) is

$$(11.1) \qquad K_0 = H_0$$

$$(11.2) \qquad K_1 = H_1 + \frac{\partial W_1}{\partial t} + \{W_1, H_0\}$$

$$(11.3) \qquad K_n = H_n + \frac{1}{n} \left[\frac{\partial W_n}{\partial t} + \{W_n, H_0\} \right] + \frac{1}{n} \sum_{m=1}^{n-1} \left[L_{n-m} \, K_m + m \, S_{n-m} \, H_m \right],$$

$$n = 2, 3, 4, \ldots$$

in which the functions H_i, K_i and W_i are written as functions of the transformed variables Q_i and P_i. By choosing W_n appropriately in eqs.(11), we may simplify the expression for K_n.

Action-angle Variables

In this Chapter we shall be interested in applying the method to systems which, when $\epsilon = 0$, correspond to N linear oscillators, cf. eq.(2). In this case all the computations are greatly simplified by using a polar coordinate system called <u>action-angle variables</u> for the $\epsilon = 0$ problem. In the case of eq.(2), if we make the (canonical) transformation from q_1, p_1 to ϕ_1, J_1 via the equations (see Exercise 3 at the end of this Chapter)

$$(12.1) \qquad q_1 = \sqrt{2 \, J_1 / \omega} \; \sin \phi_1$$

$$(12.2) \qquad p_1 = \sqrt{2 \, J_1 \, \omega} \; \cos \phi_1$$

we obtain

(13) $H(\phi_1, J_1) = \omega\, J_1$

Here the coordinate ϕ_1 is called the angle and the momentum J_1 is the action.
Note that the eq.(13) is much simpler algebraically that the equivalent eq.(2).
While the use of action-angle variables is not an essential part of the method
of Lie transforms, we shall find it useful for the systems of forced, coupled
nonlinear oscillators which we shall consider.

As an example of the method, we consider the following nonlinear
Mathieu equation [30]:

(14) $\dfrac{d^2 u}{dt^2} + (\delta + \epsilon \cos t)\, u + \epsilon\, \alpha\, u^3 = 0$

which may be written in the Hamiltonian form (1) by setting

(15) $q = u, \quad p = \dfrac{du}{dt}.$

with the Hamiltonian

(16) $H = \dfrac{p^2}{2} + (\delta + \epsilon \cos t)\, \dfrac{q^2}{2} + \epsilon\, \alpha\, \dfrac{q^4}{4}$

We shall set

(17) $\delta = \omega^2 + \delta_1\, \epsilon + \delta_2\, \epsilon^2 + \cdots$

where ω is the frequency of the $\epsilon = 0$ problem and the δ_i are detuning
parameters. The Hamiltonian may be simplified by using action-angle variables

ϕ, J of eq.(12):

$$(18) \qquad H = \omega J + \epsilon \left[\delta_1 + \cos t \right] \frac{J}{\omega} \sin^2\phi + \epsilon \, \alpha \, \frac{J^2}{\omega^2} \sin^4\phi + O(\epsilon^2)$$

The near-identity transformation of variables (10) from ϕ,J to, say, ψ,I coordinates is given by:

$$(19.1) \qquad \phi = \psi - \epsilon \, \frac{\partial W_1}{\partial I} + O(\epsilon^2)$$

$$(19.2) \qquad J = I + \epsilon \, \frac{\partial W_1}{\partial \psi} + O(\epsilon^2)$$

where we have used, e.g., $S_1\psi = L_1\psi = \{W_1,\psi\} = \dfrac{\partial W_1}{\partial \psi} \dfrac{\partial \psi}{\partial I} - \dfrac{\partial W_1}{\partial I} \dfrac{\partial \psi}{\partial \psi} = - \dfrac{\partial W_1}{\partial I}$.

In the new coordinates, the terms of the transformed Hamiltonian are:

$$(20) \qquad K_0 = \omega I$$

$$(21.1) \qquad K_1 = H_1 + \frac{\partial W_1}{\partial t} + \{W_1, H_0\}$$

$$(21.2) \qquad K_1 = \left[\delta_1 + \cos t \right] \frac{I}{\omega} \sin^2\psi + \alpha \, \frac{I^2}{\omega^2} \sin^4\psi + \frac{\partial W_1}{\partial t} + \omega \, \frac{\partial W_1}{\partial \psi}$$

In order to choose W_1 judiciously, we trigonometrically reduce the right hand side of eq.(21.2):

$$(21.3) \qquad K_1 = \frac{I}{4\omega} \left[- \cos(t+2\psi) - \cos(t-2\psi) + 2 \cos t - 2 \, \delta_1 \cos 2\psi + 2 \, \delta_1 \right]$$

$$+ \frac{\alpha I^2}{8\omega^2} \left[\cos 4\psi - 4 \cos 2\psi + 3 \right] + \frac{\partial W_1}{\partial t} + \omega \, \frac{\partial W_1}{\partial \psi}$$

Each term on the right hand side of (21.3) which is of the form

(22) $A \cos (B t + C \psi)$

may be removed by including a corresponding term in W_1 of the form:

(23) $D \sin (B t + C \psi)$

which leads to the choice

(24) $W_1 = \dfrac{I \sin(t+2\psi)}{4\omega (1+2\omega)} - \dfrac{I \sin(t-2\psi)}{4\omega (-1+2\omega)} + (\delta_1 I\omega + \alpha I^2) \dfrac{\sin 2\psi}{4\omega^3} - \dfrac{I \sin t}{2 \omega}$

$$- \frac{\alpha I^2 \sin 4\psi}{32 \omega^3}$$

and thus yields the expression

(25) $K_1 = \dfrac{\delta_1 I}{2\omega} + \dfrac{3\alpha I^2}{8\omega^2}$

Note that we do not remove the constant terms from K_1, as these would require a term in W_1 that is linear in t or ψ, and such a term would destroy the uniform validity of the asymptotic expansion for W_1 as t or ψ goes to infinity.

The significance of eq. (25) is that now the transformed Hamiltonian K does not involve ψ or t to order ϵ:

(26) $K = K_0 + K_1 \epsilon = \omega I + \epsilon \left[\dfrac{\delta_1 I}{2\omega} + \dfrac{3\alpha I^2}{8\omega^2} \right]$

Thus the transformed problem is easy to solve:

(27.1)
$$\frac{d\psi}{dt} = \frac{\partial K}{\partial I} = \omega + \frac{\delta_1 \epsilon}{2\omega} + \frac{3\alpha\epsilon}{4\omega^2} I$$

(27.2)
$$\frac{dI}{dt} = -\frac{\partial K}{\partial \psi} = 0$$

from which

(28.1)
$$\psi = \left[\omega + \frac{\delta_1 \epsilon}{2\omega} + \frac{3\alpha\epsilon}{4\omega^2} I \right] t + \psi_0$$

(28.2)
$$I = I_0$$

By substituting (28) and (24) into (19) we may obtain expressions for the original action-angle variables ϕ and J. Then an expression for the original variable $u = q = \sqrt{\frac{2J}{\omega}} \sin \phi$ is obtainable.

Computer Algebra

The key formulas (7)-(9) of the method of Lie transforms are ideally suited for implementation on MACSYMA [22]. We shall demonstrate a program called LIE on the foregoing example, and then give the program listing. We annotate the run in Italics:

LIE():
DO YOU WANT TO INPUT A NEW PROBLEM (Y/N) ?
Y;
ENTER NUMBER OF DEGREES OF FREEDOM
1;
ENTER SYMBOL FOR Q[1]
Q;
ENTER SYMBOL FOR P[1]
P;
THE HAMILTONIAN DEPENDS ON THE Q'S, P'S, T AND E (SMALL PARAMETER)

THE E=0 PROBLEM MUST BE OF THE FORM:

$$H = \frac{w_1^2 q^2 + p^2}{2}$$

ENTER THE HAMILTONIAN
P^2/2+(W^2+DEL1*E+DEL2*E^2+E*COS(T))*Q^2/2+E*A*Q^4/4;

$$H = \frac{q^2 (w^2 + e\cos(t) + del2\ e^2 + del1\ e)}{2} + \frac{a\ e\ q^4}{4} + \frac{p^2}{2}$$

THE ACTION-ANGLE VARIABLES ARE J'S FOR ACTION, PHI'S FOR ANGLE

$$H = j_1 w - \frac{j_1 e \cos(t + 2\ phi_1)}{4\ w} - \frac{j_1 e \cos(t - 2\ phi_1)}{4\ w} + \frac{j_1 e \cos(t)}{2\ w}$$

$$- \frac{j_1 \cos(2\ phi_1)\ del2\ e^2}{2\ w} + \frac{j_1\ del2\ e^2}{2\ w} - \frac{j_1 \cos(2\ phi_1)\ del1\ e}{2\ w} + \frac{j_1\ del1\ e}{2\ w}$$

$$+ \frac{j_1^2 \cos(4\ phi_1)\ a\ e}{8\ w^2} - \frac{j_1^2 \cos(2\ phi_1)\ a\ e}{2\ w^2} + \frac{3\ j_1^2\ a\ e}{8\ w^2}$$

ENTER HIGHEST ORDER TERMS IN E TO BE KEPT
1;

  The program up to this point has involved entering data and
transforming to action-angle variables j_1, phi_1. *These preliminary steps*
completed, the program begins the Lie transform computation. The transformed
action-angle variables are denoted by i_1, psi_1, *and the first result displayed*
is the generating function W_1, *referred to as WGEN[1] to avoid confusion with*
ω_1. *This equation corresponds to eq.(24) above:*

WGEN[1] =

$$\frac{i_1 \sin(t + 2 \ psi_1)}{8 w_1{}^2 + 4 w_1} - \frac{i_1 \sin(t - 2 \ psi_1)}{8 w_1{}^2 - 4 w_1} + \frac{\sin(2 \ psi_1) \ (i_1 \ dell \ w_1 + i_1 \ a_1)^2}{4 w_1{}^3}$$

$$- \frac{i_1 \sin(t)}{2 w_1} - \frac{i_1 \sin(4 \ psi_1) \ a_1{}^2}{32 w_1{}^3}$$

Having obtained W_1, *the program computes* K_1 *as in eq.(25) above:*

THE TRANSFORMED HAMILTONIAN K[1] =

$$\frac{i_1 \ dell_1}{2 w} + \frac{3 \ i_1 \ a_1{}^2}{8 w^2}$$

The foregoing steps of computing W_i *and* K_i *would now be iterated until the specified truncation order is reached. Since we specified truncation at order* ϵ, *the main loop is now exited, and the resulting expression for the transformed Hamiltonian K (called the Kamiltonian after Goldstein [12]) is displayed:*

THE TRANSFORMED HAMILTONIAN IS

$$K = i_1 \ w_1 + e \ (\frac{i_1 \ dell_1}{2 w} + \frac{3 \ i_1 \ a_1{}^2}{8 w^2})$$

The user is now given the option of generating the near-identity canonical transformation from j_1, phi_1 *to* i_1, psi_1 *action-angle coordinates:*

DO YOU WANT TO SEE THE NEAR IDENTITY TRANSFORMATION (Y/N) ?
Y;

$$j_1 = e_1 \left(\frac{2 i_1 \cos(t + 2 \text{ psi}_1)}{8 w^2 + 4 w} + \frac{2 i_1 \cos(t - 2 \text{ psi}_1)}{8 w^2 - 4 w} \right.$$

$$\left. + \frac{\cos(2 \text{ psi}_1) (i_1 \text{ dell } w + i_1 a)}{2 w^3} - \frac{i_1 \cos(4 \text{ psi}_1) a}{8 w^3} \right) + i_1$$

$$\text{phi}_1 = e_1 \left(- \frac{\sin(t + 2 \text{ psi}_1)}{8 w^2 + 4 w} + \frac{\sin(t - 2 \text{ psi}_1)}{8 w^2 - 4 w} - \frac{\sin(2 \text{ psi}_1) (\text{dell } w + 2 i_1 a)}{4 w^3} \right.$$

$$\left. + \frac{\sin(t)}{2 w} + \frac{i_1 \sin(4 \text{ psi}_1) a}{16 w^3} \right) + \text{psi}_1$$

[VAX 8500 TIME = 52 SEC.]

Note that the $O(\epsilon)$ computation would not be valid for $\omega = 1/2$ due to vanishing denominators. Such singular parameter values are called resonances. If we had extended the computation to include ϵ^2 terms, we would have found the additional resonances $\omega = 1$ and $\omega = 1/4$. This behavior is typical of nonlinear systems: as we extend the computation to higher and higher order in ϵ, we find more and more resonant parameter values.

In order to examine typical behavior at a resonant parameter value, we assign ω the value 1/4 (corresponding to 4:1 subharmonics) and again call the program LIE:

W:1/4$

LIE()$

DO YOU WANT TO INPUT A NEW PROBLEM (Y/N) ?
N;

By choosing not to input a new problem, the previous Hamiltonian is used, subject to any parameter assignments that have been made since the program was last run.

We choose to truncate at ϵ^2 terms in order to see the effect of the resonance (which does not show up at order ϵ):

ENTER HIGHEST ORDER TERMS IN E TO BE KEPT
2;
WGEN[1] =

$$
\frac{2 \text{ i} \ \sin(t + 2 \text{ psi})}{3} + 2 \text{ i} \ \sin(t - 2 \text{ psi}) - 2 \text{ i} \ \sin(t)
$$

$$
+ \sin(2 \text{ psi}) \ (4 \text{ i} \ \text{dell} + 16 \text{ i} \ a) - 2 \text{ i} \ \sin(4 \text{ psi}) \ a
$$

THE TRANSFORMED HAMILTONIAN K[1] =

$$
2 \text{ i} \ \text{dell} + 6 \text{ i} \ a
$$

WGEN[2] =

$$
\frac{4 \text{ i} \ \sin(2 \ t + 2 \text{ psi})}{15} + \frac{4 \text{ i} \ \sin(2 \ t - 2 \text{ psi})}{3} - \frac{16 \text{ i} \ a \ \sin(t + 4 \text{ psi})}{3}
$$

+ 5 additional lines omitted for brevity

THE TRANSFORMED HAMILTONIAN K[2] =

$$
8 \text{ i} \ a \ \cos(t - 4 \text{ psi}) + 2 \text{ i} \ \text{del2} - 8 \text{ i} \ \text{dell} - 96 \text{ i} \ a \ \text{dell} - 272 \text{ i} \ a
$$

$$
+ \frac{4 \text{ i}}{3}
$$

THE TRANSFORMED HAMILTONIAN IS

$$K = e_1^2 \left(8 i_1 a_1 \cos(t - 4 \psi_1) + 2 i_1 del2_1 - 8 i_1 dell_1^2 - 96 i_1 a_1 dell_1\right.$$

$$\left. - 272 i_1 a_1^3 + \frac{4 i_1}{3}\right) + (2 i_1 dell_1 + 6 i_1 a_1) e_1^2 + \frac{i_1}{4}$$

      As a check on the program, we note that this result agrees with a previously published hand calculation [30].

DO YOU WANT TO SEE THE NEAR IDENTITY TRANSFORMATION (Y/N) ?
N;

[VAX 8500 TIME = 145 SEC.]

      Note that in this resonant parameter case the transformed Hamiltonian K is not independent of angle terms as it was for nonresonant ω, cf. eq.(26). This is due to the unremovability of terms with vanishing denominators. Nevertheless by making the canonical transformation from ψ_1, I_1 to $\tilde{\psi}, \tilde{I}$ via

(29) $$\tilde{\psi} = \psi_1 - t/4 , \quad \tilde{I} = I_1$$

the new Hamiltonian becomes autonomous, and is therefore relatively easy to analyze. See Exercise 1 at the end of this Chapter.

      Here is the program listing:

```
LIE():=BLOCK(
/* INPUT PROBLEM ? */
CHOICE1:READ("DO YOU WANT TO INPUT A NEW PROBLEM (Y/N) ?"),
IF CHOICE1=N THEN GO(JUMP1),
/* INPUT PROBLEM */
N:READ("ENTER NUMBER OF DEGREES OF FREEDOM"),
```

```
FOR II:1 THRU N DO (

   Q[II]:READ("ENTER SYMBOL FOR Q[",II,"]"),

   P[II]:READ("ENTER SYMBOL FOR P[",II,"]")),

KILL(W),

PRINT("THE HAMILTONIAN DEPENDS ON THE Q'S, P'S, T AND E (SMALL PARAMETER)"),

PRINT("THE E=0 PROBLEM MUST BE OF THE FORM:"),

PRINT("H =",SUM((P[II]^2+W[II]^2*Q[II]^2)/2,II,1,N)),

HORIGINAL:READ("ENTER THE HAMILTONIAN"),

PRINT("H =",HORIGINAL),

/* TRANSFORM TO ACTION-ANGLE VARIABLES */

/* FIND THE W[II]'S */

HO:EV(HORIGINAL,E=0),

FOR II:1 THRU N DO

   W[II]:SQRT(DIFF(HO,Q[II],2)),

PRINT("THE ACTION-ANGLE VARIABLES ARE J'S FOR ACTION, PHI'S FOR ANGLE"),

FOR II:1 THRU N DO

   TR[II]:[Q[II]=SQRT(2*J[II]/W[II])*SIN(PHI[II]),

           P[II]=SQRT(2*J[II]*W[II])*COS(PHI[II])],

TRAN:MAKELIST(TR[II],II,1,N),

H:EV(HORIGINAL,TRAN,ASSUME_POS:TRUE,INFEVAL),

H:TRIGSIMP(H),

H:EXPAND(TRIGREDUCE(EXPAND(H))),

PRINT("H =",H),


JUMP1,

/* INPUT TRUNCATION ORDER */

NTRUNC:READ("ENTER HIGHEST ORDER TERMS IN E TO BE KEPT"),

FOR II:0 THRU NTRUNC DO

   H[II]:RATCOEF(H,E,II),
```

```
/* LIE TRANSFORMS */

/* NEAR IDENTITY TRANSFORMATION FROM (J,PHI)'S TO (I,PSI)'S */

/* UPDATE VARIABLES */

FOR II:1 THRU N DO(

    P[II]:I[II],

    Q[II]:PSI[II]),

/* REPLACE J AND PHI BY I AND PSI IN H'S */

FOR II:0 THRU NTRUNC DO

    H[II]:EV(H[II],MAKELIST(J[III]=I[III],III,1,N),

                 MAKELIST(PHI[III]=PSI[III],III,1,N)),

K[0]:H[0],

/* DECLARE WGEN[I] TO BE A FN OF T, Q'S AND P'S */

KILL(WGEN),

DEPENDS(WGEN,[T]),

FOR II:1 THRU N DO

  DEPENDS(WGEN,[Q[II],P[II]]),

/* E=0 PROBLEM IS OF FORM SUM(W[II]*I[II]) */

/* CHOOSE WGEN[II] TO KILL AS MUCH AS POSSIBLE IN EQ(II) */

/* EQUATE K[II] TO UNREMOVABLE TERMS */

/* DEFINE PATTERN MATCHING RULES TO ISOLATE ARGS OF TRIG TERMS */

MATCHDECLARE([DUMMY1,DUMMY2],TRUE),

DEFRULE(COSARG,DUMMY1*COS(DUMMY2),DUMMY2),

DEFRULE(SINARG,DUMMY1*SIN(DUMMY2),DUMMY2),

FOR II:1 THRU NTRUNC DO(

    EQN[II]:EXPAND(TRIGREDUCE(EXPAND(EQ(II)))),

    TEMP:EXPAND(EV(EQN[II],WGEN[II]=0)),

/* CHANGE SUM TO A LIST */

    TEMP1:ARGS(TEMP),
```

```
/* REMOVE CONSTANT TERMS */

   TEMP2:MAP(TRIGIDENTIFY,TEMP1),

/* ISOLATE ARGUMENTS OF TRIG TERMS */

   ARG1:APPLY1(TEMP2,COSARG,SINARG),

/* REMOVE DUPLICATE ARGUMENTS */

   ARG2:SETIFY(ARG1),

/* REMOVE RESONANT ARGUMENTS */

   ARG3:SUBLIST(ARG2,NOTRESP),

/* CHOOSE WGEN TO ELIMINATE NONRESONANT TERMS */

   LENG:LENGTH(ARG3),

   WGENTEMP1:0,

   FOR JJ:1 THRU LENG DO(

       WGENTEMP2:AAA*COS(PART(ARG3,JJ))+BBB*SIN(PART(ARG3,JJ)),

       TEMP4:EV(EQN[II],WGEN[II]=WGENTEMP2,DIFF),

       TEMP5:SOLVE([RATCOEF(TEMP4,COS(PART(ARG3,JJ))),

                    RATCOEF(TEMP4,SIN(PART(ARG3,JJ)))],[AAA,BBB]),

       WGENTEMP1:WGENTEMP1+EV(WGENTEMP2,TEMP5)),

   WGEN[II]:WGENTEMP1,

   PRINT("WGEN[",II,"] ="),

   PRINT(WGEN[II]),

   K[II]:EXPAND(EV(EQN[II],DIFF)),

   K[II]:EXPAND(RATSIMP(K[II])),

   PRINT("THE TRANSFORMED HAMILTONIAN K[",II,"] ="),

   PRINT(K[II])),

KAMILTONIAN:SUM(K[II]*E^II,II,0,NTRUNC),

PRINT ("THE TRANSFORMED HAMILTONIAN IS "),

PRINT ("K =",KAMILTONIAN),

CHOICE2:READ("DO YOU WANT TO SEE THE NEAR IDENTITY TRANSFORMATION (Y/N) ?"),

IF CHOICE2=N THEN GO(END),
```

```
/* THE NEAR IDENTITY TRANSFORMATION */
FOR II:1 THRU N DO(
    JTRANS[II]:SUM(S(III,P[II])*E^III,III,0,NTRUNC),
    PHITRANS[II]:SUM(S(III,Q[II])*E^III,III,0,NTRUNC)),
FOR II:1 THRU N DO(
    PRINT(J[II],"=",JTRANS[II]),
    PRINT(PHI[II],"=",PHITRANS[II])),

END,

KAMILTONIAN)$
```

```
/* AUXILIARY FUNCTIONS */
POISSON(F,G):=
        SUM(DIFF(F,Q[II])*DIFF(G,P[II])-DIFF(F,P[II])*DIFF(G,Q[II]),II,1,N)$


L(II,F):=POISSON(WGEN[II],F)$


S(II,F):=(IF II=0 THEN F ELSE SUM(L(II-M,S(M,F)),M,0,II-1)/II)$


EQ(II):=(H[II]+(DIFF(WGEN[II],T)+POISSON(WGEN[II],H[0])))/II
        +SUM(L(II-M,K[M])+M*S(II-M,H[M]),M,1,II-1)/II)$


LZAP(ANY):=DIFF(ANY,T)+POISSON(ANY,H[0])$


TRIGIDENTIFY(EXP):=IF FREEOF(SIN,EXP) AND FREEOF(COS,EXP) THEN 0 ELSE EXP$


NOTRESP(EXP):=NOT IS(LZAP(EXP) = 0)$
```

```
SETIFY(LIST):=(

    SET:[LIST[1]],

    FOR I:2 THRU LENGTH(LIST) DO(

        IF NOT MEMBER(LIST[I],SET) THEN SET:CONS(LIST[I],SET)),

    SET)$
```

Example with Two Degrees of Freedom

In order to illustrate a two degree of freedom example, we take the problem of two identical nonlinear oscillators with nonlinear coupling [28],[29]:

$$(30) \qquad H = \frac{p_1^2}{2} + \frac{q_1^2}{2} + \frac{p_2^2}{2} + \frac{q_2^2}{2} + \epsilon \frac{k}{4} \left[q_1^4 + q_2^4\right] + \frac{\epsilon}{4} \left[q_1 - q_2\right]^4$$

This problem involves a resonance associated with the fact that the frequencies of the uncoupled linear oscillators are equal. Here is the MACSYMA run of the program LIE applied to this example:

```
LIE();
DO YOU WANT TO INPUT A NEW PROBLEM (Y/N) ?
Y;
ENTER NUMBER OF DEGREES OF FREEDOM
2;
ENTER SYMBOL FOR Q[ 1 ]
Q1;
ENTER SYMBOL FOR P[ 1 ]
P1;
ENTER SYMBOL FOR Q[ 2 ]
Q2;
ENTER SYMBOL FOR P[ 2 ]
P2;
THE HAMILTONIAN DEPENDS ON THE Q'S, P'S, T AND E (SMALL PARAMETER)
THE E=0 PROBLEM MUST BE OF THE FORM:

      2   2     2      2   2     2
    w  q2  + p2      w  q1  + p1
    2                1
H = --------------- + -------------
          2                 2
```

ENTER THE HAMILTONIAN
P1^2/2+Q1^2/2+P2^2/2+Q2^2/2+E*K*(Q1^4/4+Q2^4/4)+E*(Q1-Q2)^4/4;

$$
H = e\,k \left(\frac{q2^4}{4} + \frac{q1^4}{4}\right) + \frac{q2^2}{2} + \frac{e\,(q1 - q2)^4}{4} + \frac{q1^2}{2} + \frac{p2^2}{2} + \frac{p1}{2}
$$

THE ACTION-ANGLE VARIABLES ARE J'S FOR ACTION, PHI'S FOR ANGLE

$$
H = \frac{j_2^2\,\cos(4\,phi_2)\,e\,k}{8} - \frac{j_2^2\,\cos(2\,phi_2)\,e\,k}{2} + \frac{j_1^2\,\cos(4\,phi_1)\,e\,k}{8}
$$

+ 8 additional lines omitted for brevity

$$
+ \frac{3\,j_1\,j_2\,e}{2} + \frac{3\,j_1^2\,e}{8} + j_2 + j_1
$$

ENTER HIGHEST ORDER TERMS IN E TO BE KEPT
1;

WGEN[1] =

$$
\frac{\sin(2\,psi_2)\,(i_2\,k + i_2^2 + 3\,i_1\,i_2)}{4} - \frac{\sin(4\,psi_2)\,(i_2\,k + i_2^2)}{32}
$$

+ 5 additional lines omitted for brevity

THE TRANSFORMED HAMILTONIAN K[1] =

$$
\frac{3\,i_2\,k}{8} + \frac{3\,i_1\,k}{8} + \frac{3\,i_1\,i_2\,\cos(2\,psi_2 - 2\,psi_1)}{4}
$$

$$
- \frac{3\,\sqrt{i_1}\,i_2^{3/2}\,\cos(psi_2 - psi_1)}{2} - \frac{3\,i_1^{3/2}\,\sqrt{i_2}\,\cos(psi_2 - psi_1)}{2} + \frac{3\,i_2^2}{8}
$$

$$
+ \frac{3\,i_1\,i_2}{2} + \frac{3\,i_1^2}{8}
$$

THE TRANSFORMED HAMILTONIAN IS

```
        2          2
   3 i  k     3 i  k     3 i  i  cos(2 psi  - 2 psi )
        2          1         1 2          2        1
K = e (------- + ------- + -----------------------------
        8           8                    4
```

```
        3/2                               3/2
   3 sqrt(i ) i    cos(psi  - psi )   3 i     sqrt(i ) cos(psi  - psi )   3 i
          1   2         2      1         1          2          2      1         2
 - ---------------------------------- - ---------------------------------- + ----
                   2                                    2                      8
```

```
        2
   3 i  i    3 i
      1  2      1
 + ------- + ----) + i  + i
      2        8       2    1
```

DO YOU WANT TO SEE THE NEAR IDENTITY TRANSFORMATION (Y/N) ?
N;

[VAX 8500 TIME = 181 SEC.]

The resulting Kamiltonian,

$$(31) \qquad K = I_1 + I_2 + \epsilon \left[\frac{3}{8} (k+1) (I_1^2 + I_2^2) + \frac{3}{2} I_1 I_2 \right.$$

$$\left. - \frac{3}{2} \sqrt{I_1 I_2} (I_1 + I_2) \cos(\psi_2 - \psi_1) + \frac{3}{4} I_1 I_2 \cos 2(\psi_2 - \psi_1) \right]$$

can be simplified by the canonical transformation:

$$(32) \qquad \tilde{\psi}_1 = \psi_1 - \psi_2, \ \tilde{I}_1 = I_1, \ \tilde{\psi}_2 = \psi_2, \ \tilde{I}_2 = I_1 + I_2$$

The resulting expression for K in terms of $\tilde{\psi}_1, \tilde{I}_1, \tilde{\psi}_2, \tilde{I}_2$ does not involve $\tilde{\psi}_2$.
Thus $\tilde{I}_2$ is a constant in time (since $\partial K / \partial \tilde{\psi}_2 = 0$), and to $O(\epsilon)$ the problem can be
reduced to that of a single degree of freedom system.

Exercises

1. For eq.(14) with $\omega = 1/4$ in eq.(17), show that the Hamiltonian K which we
obtained,

$$K = \frac{I_1}{4} + \epsilon \left[2\delta_1 I_1 + 6\alpha I_1^2\right]$$

$$+ \epsilon^2 \left[8\alpha I_1^2 \cos(t-4\psi_1) + 2\delta_2 I_1 - 8\delta_1^2 I_1 - 96\alpha\delta_1 I_1^2 - 272\alpha^2 I_1^3 + \frac{4}{3}I_1\right]$$

gives equations of motion which are equivalent to those corresponding to the
new Hamiltonian $\tilde{K}(\tilde{\psi}, \tilde{I})$ under the transformation (29):

(P1) $$\tilde{K} = \epsilon \left[2\delta_1 \tilde{I} + 6\alpha \tilde{I}^2\right]$$

$$+ \epsilon^2 \left[8\alpha \tilde{I}^2 \cos 4\tilde{\psi} + 2\delta_2 \tilde{I} - 8\delta_1^2 \tilde{I} - 96\alpha\delta_1 \tilde{I}^2 - 272\alpha^2 \tilde{I}^3 + \frac{4}{3}\tilde{I}\right]$$

The equilibria of the corresponding equations of motion represent 4:1
subharmonic periodic motions of the original system (from eqs.(29),(12)).
Obtain an asymptotic expansion for the location of the equilibria of this
system, neglecting terms of $O(\epsilon^2)$. Using the fact that $\tilde{I}$ must be non-negative,
show that there exists a bifurcation curve in the δ-ϵ plane given by:

(P2) $$\delta = \frac{1}{16} - \frac{2}{3}\epsilon^2 + \cdots$$

such that points lying to the left of this curve exhibit two 4:1 subharmonics,
while points lying to the right exhibit no such subharmonics [30].

2. Consider eq.(14) with $\omega = 1/2$ in eq.(17). From eq.(24) this choice of ω leads to a resonance in the order ϵ terms. Investigate the equilibria in the resulting Kamiltonian equations. Show that there are two bifurcation curves in the δ-ϵ plane given by:

(P3)
$$\delta = \frac{1}{4} \pm \frac{1}{2} \epsilon + \cdots$$

which separate regions with 0, 2, and 4 nontrivial equilibria. Show that these respectively correspond to 0, 1, and 2 2:1 subharmonic periodic motions [18],[30].

3. A transformation is said to be canonical if it preserves the Hamiltonian structure of the differential equations of motion. The conditions for a transformation from (q_i, p_i) to (Q_i, P_i) variables to be canonical is [3]:

(P4)
$$\sum dq_i {}^{\wedge} dp_i = \sum dQ_i {}^{\wedge} dP_i$$

where the wedge product $^{\wedge}$ is anticommutative:

(P5)
$$df {}^{\wedge} dg = - dg {}^{\wedge} df, \quad df {}^{\wedge} df = 0$$

In computing the differentials dq_i and dp_i in (P4), we use the ordinary chain rule based on the transformation of variables. This gives dq_i and dp_i as linear combinations of the dQ_i's and dP_i's. Show that the transformations (12), (29) and (32) used in this Chapter are canonical.

4. Use Lie transforms to treat the undamped Duffing equation:

(P6) $$\frac{d^2x}{dt^2} + (1 + \epsilon \Delta) x + \epsilon \alpha x^3 = \epsilon A \cos t$$

Take $q = x$ and $p = \frac{dx}{dt}$ and use the Hamiltonian:

(P7) $$H = \frac{p^2}{2} + (1 + \epsilon \Delta) \frac{q^2}{2} + \epsilon \alpha \frac{q^4}{4} - \epsilon A q \cos t$$

Neglecting terms of $O(\epsilon^2)$, find an approximate relationship between the amplitude of the periodic response and the parameters A, α and Δ.

CHAPTER 7

LIAPUNOV-SCHMIDT REDUCTION

Introduction

Like the center manifold reduction, the Liapunov-Schmidt reduction is a
method which replaces a large and complicated set of equations by a simpler and
smaller system which contains all the essential information concerning a
bifurcation. The method is applicable to a system of nonlinear evolution
equations of the form

$$(1) \qquad \frac{dy}{dt} = F(y,\alpha)$$

where α is a parameter. A solution to (1), $y = y(t,\alpha)$, is said to exhibit a
bifurcation as α passes through a critical value α_c, if the nature of the
solutions exhibits a qualitative change at α_c, e.g. in the case of an
equilibrium solution, if the number of equilibria or their type of stability
changes. Here (1) describes either a finite dimensional or infinite
dimensional system. The former we may think of a system of ordinary
differential equations, whereas the latter would be represented by one or more
partial differential equations or integrodifferential equations.

The aim of the Liapunov-Schmidt reduction is to reduce (1), which, even
if finite dimensional, may be a very large system, to a small system of
algebraic equations (typically one or two dimensional). This will be

accomplished by means of a perturbation expansion near a bifurcation point of a known solution of (1). It can be shown [13],[42] that solutions of these algebraic equations are locally in one to one correspondence to the solutions of (1). An important restriction for our programs will be that we confine ourselves to a <u>nondegenerate bifurcation from a stationary solution to another stationary solution</u>. This excludes the important case of Hopf bifurcations, which in principle one can treat with similar methods (cf.[13]). Thus we shall be interested in steady state solutions to the system (1):

$$F(y,\alpha) = 0.$$

Example: Euler Buckling

In order to introduce the method we perform an ad hoc perturbation analysis of the buckling problem. We will generalize our approach in the next section. A model for static beam buckling under compression (Euler's elastica, [25]) is given by

(2) $$\frac{d^2y}{dx^2} + \alpha \sin y = 0 , \qquad y'(0) = y'(1) = 0$$

where x is the arc length along the deformed beam, y(x) is the angle which the tangent to the beam makes with the undeformed beam axis, α is the load parameter and where the beam's length is normalized to 1. The boundary conditions correspond to pinned ends. Eq.(2) has an analytical solution in terms of elliptic integrals (cf.[40])

(3) $$\alpha = 4\, m^2\, K^2(k) \qquad m = 1,2,3,\ldots$$

where $K(k) = \int_0^{\pi/2} (1-k^2\sin^2\varphi)^{-1/2} \, d\varphi$ is the complete elliptic integral of the first kind and $k = \sin a/2$ where $a = y(0)$ is the beam angle at $x = 0$.

If we linearize (2) we have the equation for an harmonic oscillator

$$(4) \qquad \frac{d^2y}{dx^2} + \alpha y = 0 , \qquad y'(0) = y'(1) = 0$$

Note that $y \equiv 0$ is the only solution satisfying the boundary conditions if $\alpha \neq n^2\pi^2$, $n = 1,2,3,\cdots$. If $\alpha = n^2\pi^2$ then $y = A \cos n\pi x$ is an admissible solution with undetermined amplitude A. Let us assume we increase α from 0. Then $\alpha = \alpha_c = \pi^2$ is the first value where a nontrivial solution may branch off from the solution $y \equiv 0$. In order to find the deflection amplitude A as a function of a parameter $\lambda = \alpha - \alpha_c$, we make the following ansatz:

$$(5) \qquad y = A \cos \pi x + w(x;A,\lambda)$$

where A is assumed to be small and where w and λ are of second order in A. We require that a Fourier decomposition of $w(x;A,\lambda)$ has no component $\cos \pi x$, since if it had we could include it in the $A \cos \pi x$ term. Expanding (2) up to third order we find

$$\frac{d^2y}{dx^2} + \alpha \left(y - \frac{y^3}{3!} \right) =$$

$$\frac{d^2}{dx^2} (A \cos \pi x + w) + (\lambda + \pi^2) \left[A \cos \pi x + w - \frac{(A \cos \pi x + w)^3}{3!} \right] =$$

$$(6) \qquad \lambda A \cos \pi x + \frac{d^2w}{dx^2} + (\lambda + \pi^2) \left[w - \frac{(A \cos \pi x + w)^3}{3!} \right] = 0$$

Since {cos nπx} is a complete orthogonal set on [0,1], we can split the Fourier decomposition of (6) into two parts: The cos πx component and all other modes. To obtain the coefficient of the cos πx component of (6), we take the inner product of (6) with cos πx and divide it by $\int_0^1 \cos^2 \pi x \, dx = 1/2$. Since w, and hence w'', have, by hypothesis, no cos πx components, we obtain (after some trigonometric simplification):

(7.1) $\lambda A - \dfrac{(\lambda + \pi^2)}{6} \Bigg[\dfrac{3}{4} A^3 +$

$\int_0^1 \Bigg[\dfrac{3}{2} A^2 w \cos 3\pi x + 3A w^2 (1 + \cos 2\pi x) + 2w^3 \cos \pi x \Bigg] dx \Bigg] = 0$

All the other modes may be accounted for by subtracting the cos πx component (the coefficient of which is given by Eq(7.1)) from (6):

(7.2) $\dfrac{d^2 w}{dx^2} + (\lambda + \pi^2) w$

$- \dfrac{(\lambda + \pi^2)}{6} \Bigg[\dfrac{1}{4} A^3 \cos 3\pi x + \dfrac{3}{2} A^2 w (1 + \cos 2\pi x) + 3A w^2 \cos \pi x + w^3$

$- \cos \pi x \int_0^1 \Bigg[\dfrac{3}{2} A^2 w \cos 3\pi x + 3A w^2 (1 + \cos 2\pi x) + 2w^3 \cos \pi x \Bigg] dx \Bigg] = 0$

We will later recognize eq. (7.1) as the <u>bifurcation equation</u>. Note that to solve (7.1) we have to know w and hence we must first solve (7.2). A closer look at (7.2) reveals that we are dealing with a forced nonlinear detuned oscillator. Since w is assumed to be of second order in A and λ, eq.(7.2) reduces, for small A and λ, to a linear forced harmonic oscillator

(8) $$\frac{d^2w}{dx^2} + \pi^2 w - \frac{\pi^2 A^3}{24} \cos 3\pi x = 0$$

The solution of the unforced system is given by $w = B \cos \pi x + C \sin \pi x$. A particular solution to (8) is given by

(9) $$w = - \frac{A^3}{192} \cos 3\pi x$$

and hence, in order to fit the boundary conditions $w'(0) = w'(1) = 0$, we must set $C \equiv 0$. Morever B must be zero since w must not have a Fourier component $\cos \pi x$. Thus (9) represents the solution w of (7.2). Therefore up to order A^3, the integrated terms in eq.(7.1) are negligible and we calculate the response function $A(\lambda)$ to be a solution of

$$\lambda A - \frac{\pi^2}{8} A^3 = 0$$

However, if we include terms in eq.(2) up to order A^5, and then determine the Fourier coefficient of the $\cos \pi x$ mode and insert (9) into the resulting equation, we find:

(10) $$\lambda A - \frac{\pi^2}{8} A^3 - \frac{\lambda}{8} A^3 + \frac{3 \pi^2}{512} A^5 = 0$$

We solve (10) for λ and expand the result up to fourth order:

(11) $$\lambda = \pi^2 \left(\frac{1}{8} A^2 + \frac{5}{512} A^4 \right)$$

A Taylor expansion of the elliptic integral in the exact solution (3) with respect to a, the beam angle at $x = 0$, gives

$$(12) \qquad \alpha = \pi^2 \left(1 + \frac{1}{8} a^2 + \frac{17}{1536} a^4\right)$$

If we take into account that $a = y(0) = A + w(0;A,0) = A - \dfrac{A^3}{192} + \cdots$, and that $\alpha = \lambda + \pi^2$, then (11) and (12) are identical up to order 4.

Formal Setup

In the previous example we made several assumptions about the size of small terms, e.g. w and λ were assumed to be $O(A^2)$. These assumptions are, as we will show, both unnecessary and, from the point of view of an automated computer algebra program, undesirable. In this section we shall generalize the perturbation analysis using the language of functional analysis in order to prepare for the approach used in the MACSYMA program. We assume a physical system whose dynamics are governed by the evolution equation (1), where y is an element of an appropriate Banach space E and $y \equiv 0$ is a trivial solution. E is typically a subspace of the Hilbert space $L^2(\Omega)$ where Ω is a bounded domain in $\mathbb{R}^n$, furnished with the standard inner product

$$(13) \qquad \langle u,v \rangle = \int_\Omega u(x) \cdot v(x) \ dx$$

where $u(x)$ and $v(x)$ are n-vectors whose elements are functions of x, and where $u(x) \cdot v(x)$ is the scalar product of u and v in $\mathbb{R}^n$.

If $D_y F(0,\alpha)$, the linear part of the operator F (i.e. its Frechet derivative), has only eigenvalues with negative real parts, then every perturbation of $y \equiv 0$ in (1) is damped and will die away. However if we have an eigenvalue of L which is zero at $\alpha = \alpha_c$, then the situation is marginal and signifies the onset of branching at the bifurcation point $\lambda \equiv \alpha - \alpha_c = 0$. We define the linear operator L to be $D_y F(0,\alpha_c)$. Since we only want to deal with nondegenerate bifurcations, the eigenvalues μ_n and the eigenfunctions y_n of L will satisfy $Ly_1 = 0$ and $Ly_n = \mu_n y_n$ where Re $\mu_n < 0$. We call y_1 the critical eigenfunction and define the adjoint critical eigenfunction via $L^* y_1^* = 0$, where L^* is the adjoint operator of L, i.e., $\langle L^* f, g \rangle = \langle f, Lg \rangle$. We assume that we can decompose the Banach space $E = N + R$ into kernel N and range R of L at $\lambda = 0$ such that we can write $y = A y_1 + w$, where $A \in \mathbb{R}$ is a "state variable" or "amplitude" and $y_1 \in N$, $w \in R$. Thus $Ly_1 = 0$ and $w = Lu$ for some u. In all our examples this splitting is always possible. Let us define projections onto the subspaces N and R:

(14.1)
$$P y = \frac{\langle y, y_1^* \rangle}{\langle y_1, y_1^* \rangle} y_1$$

(14.2)
$$Q = Id - P, \text{ i.e., } Q y = y - \frac{\langle y, y_1^* \rangle}{\langle y_1, y_1^* \rangle} y_1$$

P projects onto the kernel N of L and Q onto the image R of L. Observe that $P y_1 = y_1$ and $Q y_1 = 0$. Since by definition $L^* y_1^* = 0$ and since w is an element of the range of L we find that $\langle L^* y_1^*, u \rangle = \langle y_1^*, Lu \rangle = \langle y_1^*, w \rangle = 0$. Hence from (14.2) we find that $Q w = w$. Next we replace the parameter α by $\lambda = \alpha - \alpha_c$ in $F(y, \alpha) = 0$, and we write it in the form

(15)
$$F(y, \lambda) = L y + \tilde{F}(y, \lambda) = 0$$

Thus we can split (15) into two equations on the subspaces N and R:

(16.1) $P F = P \tilde{F}(A y_1 + w, \lambda) \equiv \dfrac{g(A, \lambda, w)}{\langle y_1, y_1^* \rangle} y_1 = 0$

(16.2) $Q F = Q [L (A y_1 + w) + \tilde{F}(A y_1 + w, \lambda)] = L w + Q \tilde{F}(A y_1 + w, \lambda) = 0$

By this splitting we restrict L to operate on the complement of the kernel of L
and hence L is invertible on R. Therefore we can find a unique solution
$w(x; A, \lambda)$ to (16.2). Inserting this into (16.1) we get a simple algebraic
equation $g(A, \lambda, w(x; A, \lambda)) = 0$ (or simply $g(A, \lambda) = 0$) which is the so-called
bifurcation equation. In our previous example, eq.(7.1) corresponds to the
projection $P F = 0$ and eq.(7.2) to $Q F = 0$. Since $w(x; A, \lambda)$ can rarely be found
analytically in closed form as a solution of (16.2), one determines $w(x; A, \lambda)$ as
a Taylor series in A and λ to a fixed order. This leads consequently to a
truncated Taylor series for $g(A, \lambda)$. Note that in order to determine $g(A, \lambda)$ to
order n, we need $w(x; A, \lambda)$ only to order n-1, since w enters (16.1) nonlinearly
in y and λ. This gives us the nice possiblity of determining g to order n
without solving (16.2) if we know a priori (e.g. from symmetry arguments) that
w is of order n. We illustrate the method by reworking the elastica problem.

Euler Buckling Revisited

In this section we will redo the elastica problem in the way abstractly
described in the previous section. The method we follow here will be used in
our computer algebra implementation of the Liapunov-Schmidt reduction. Let us
recall the differential equation for beam buckling

(17) $\dfrac{d^2 y}{dx^2} + \alpha \sin y = 0$, $y'(0) = y'(1) = 0$

Based on our previous analysis, we take $y_1 = \cos \pi x$ as critical eigenfunction and $y_1^* = y_1$ as adjoint critical eigenfunction at $\alpha = \alpha_c = \pi^2$. We define $\lambda = \alpha - \pi^2$ and substitute $y = A \cos \pi x + w(x;A,\lambda)$ into (17) to give

$$(18) \qquad - \pi^2 A \cos \pi x + \frac{d^2 w}{dx^2} + (\lambda + \pi^2) \sin(A \cos \pi x + w) = 0$$

Our linear operator L is given by $\frac{d^2}{dx^2} + \pi^2$ and since we bifurcate from the trivial solution $y = 0$ we have

$$(19) \qquad y\big|_{A=\lambda=0} = w(x;0,0) = 0$$

Since $g(A,\lambda)$ and $w(x;A,\lambda)$ are polynomials in A and λ, we can calculate them via Taylor expansions of equations (16). The bifurcation equation $g(A,\lambda)$ of (16.1) is obtained by taking the inner product of eq.(18) with y_1^*:

$$(20) \quad g(A,\lambda) = \int_0^1 \cos \pi x \left[- \pi^2 A \cos \pi x + \frac{d^2 w}{dx^2} + (\lambda + \pi^2) \sin(A \cos \pi x + w) \right] dx$$

$$= - \frac{\pi^2 A}{2} + \int_0^1 \cos \pi x \left[\frac{d^2 w}{dx^2} + (\lambda + \pi^2) \sin(A \cos \pi x + w) \right] dx$$

Eq.(16.2) is obtained by operating with Q on (18), i.e., by subtracting $g(A,\lambda) \, y_1/\langle y_1, y_1^* \rangle = 2 \, g(A,\lambda) \cos \pi x$ from (18) (cf. (14.2)), which gives

$$(21) \qquad \frac{d^2 w}{dx^2} + (\lambda + \pi^2) \sin(A \cos \pi x + w)$$

$$- 2 \cos \pi x \int_0^1 \cos \pi x \left[\frac{d^2 w}{dx^2} + (\lambda + \pi^2) \sin(A \cos \pi x + w) \right] dx = 0$$

We will now obtain the first few terms of the Taylor expansion of $g(A,\lambda)$. We use subscripts to represent the partial derivatives of g and w evaluated at $A = 0$ and $\lambda = 0$, e.g. $g_\lambda = \dfrac{dg}{d\lambda}\bigg|_{A=\lambda=0}$. We begin by setting $A = \lambda = 0$ in (20) and using (19) to obtain:

$$(22) \qquad\qquad\qquad g(0,0) = 0$$

Next we differentiate (20) with respect to λ

$$\frac{dg}{d\lambda} = \int_0^1 \cos \pi x \left[\frac{d^3 w}{dx^2 d\lambda} + \sin(A \cos \pi x + w) + (\lambda + \pi^2)\, \frac{dw}{d\lambda} \cos(A \cos \pi x + w) \right] dx,$$

set $A = \lambda = 0$, and use (19) to obtain

$$(23) \qquad g_\lambda = \int_0^1 \cos \pi x \left[\frac{d^3 w}{dx^2 d\lambda} + \pi^2 \frac{dw}{d\lambda} \right] dx \,\Bigg|_{A=\lambda=0} = 0$$

The last equality holds since $\dfrac{d^2}{dx^2} \dfrac{dw}{d\lambda} + \pi^2 \dfrac{dw}{d\lambda} = L \dfrac{dw}{d\lambda}$ and

$\langle y_1^*, L \dfrac{dw}{d\lambda} \rangle = \langle L^* y_1^*, \dfrac{dw}{d\lambda} \rangle = 0$ since $L^* y_1^* = 0$. This result may be obtained directly from (23) by using integration by parts.

Next we differentiate (20) with respect to A,

$$\frac{dg}{dA} = -\frac{\pi^2}{2} + \int_0^1 \cos \pi x \left[\frac{d^3 w}{dx^2 dA} + (\lambda + \pi^2)\,(\cos \pi x + \frac{dw}{dA}) \cos(A \cos \pi x + w) \right] dx,$$

set $A = \lambda = 0$, and use (19) to obtain

(24) $g_A = -\dfrac{\pi^2}{2} + \displaystyle\int_0^1 \cos \pi x \left[\dfrac{d^3 w}{dx^2 dA} + \pi^2 \cos \pi x + \pi^2 \dfrac{dw}{dA} \right] dx \Bigg|_{A=\lambda=0}$

$= \displaystyle\int_0^1 \cos \pi x \left[\dfrac{d^3 w}{dx^2 dA} + \pi^2 \dfrac{dw}{dA} \right] dx \Bigg|_{A=\lambda=0} = 0$

The last equality holds by the same reasoning as used in deriving (23).

Next we take the mixed partial derivative of (20) with respect to A and λ, set A = λ = 0, and use (19) to obtain

(25) $g_{A\lambda} = \displaystyle\int_0^1 \cos \pi x \left[\dfrac{d^2}{dx^2} w_{A\lambda} + \cos \pi x + w_A + \pi^2 w_{A\lambda} \right] dx$

$= \displaystyle\int_0^1 \cos \pi x \left[\cos \pi x + w_A \right] dx$

$= \dfrac{1}{2} + \displaystyle\int_0^1 \cos \pi x \ w_A \ dx$

where we have again used $L^* y_1^* = 0$. We see that in order to find $g_{A\lambda}$ we need to know w_A. This is accomplished via eq.(21), the Q-projection of eq.(18). We differentiate (21) with respect to A, set A = λ = 0, and use (19) to obtain

(26) $\dfrac{d^2}{dx^2} w_A + \pi^2 (\cos \pi x + w_A)$

$- 2 \cos \pi x \displaystyle\int_0^1 \cos \pi x \left[\dfrac{d^2}{dx^2} w_A + \pi^2 (\cos \pi x + w_A) \right] dx = 0$

The integral in (26) can again be simplified by integrating by parts and using $L^* y_1^* = 0$, giving the simple result

(27)
$$\frac{d^2}{dx^2} w_A + \pi^2 w_A = 0$$

Hence we have to solve $L w_A = 0$ for w_A. However since $w \in R$ and since L is invertible on R, we can conclude that

(28)
$$w_A = 0$$

For our specific case this can be seen by noting that $L q = \frac{d^2 q}{dx^2} + \pi^2 q = 0$ with $q'(0) = q'(1) = 0$ has the solution $q = \cos \pi x$, which is the critical eigenfunction y_1. Since we defined w to lie in the complement of the critical eigenfunction, we arrive at the desired conclusion. Therefore, from (25),

(29)
$$g_{A\lambda} = \frac{1}{2}$$

In this manner one can determine g as a polynomial in A and λ up to any desired order. It turns out that the first nontrivial coefficient for w is given by w_{AAA}. To find it, we differentiate (21) three times with respect to A, set $A = \lambda = 0$, and use (19) and (28) to obtain

(30)
$$\frac{d^2}{dx^2} w_{AAA} + \pi^2 w_{AAA} - \pi^2 \cos^3 \pi x + 2 \cos \pi x \int_0^1 \pi^2 \cos^4 \pi x \, dx = 0$$

which can be more concisely written as

(31)
$$L w_{AAA} - \pi^2 Q(\cos^3 \pi x) = 0$$

We write $\cos^3 \pi x$ as $\frac{1}{4} (\cos 3\pi x + 3 \cos \pi x)$ and note that for this bifurcation problem the projection Q can be performed by simply deleting the critical eigenmode in the nonhomogeneous part of (31). This leaves us with

$L\ w_{AAA} = \dfrac{\pi^2}{4}\ \cos 3\pi x$. Taking into account that the homogeneous equation has

zero as a solution in R, we find that

$$(32) \qquad\qquad w_{AAA} = -\ \frac{1}{32}\ \cos 3\pi x$$

Computer Algebra

The MACSYMA program REDUCTION1 which we present in this section applies the Liapunov–Schmidt method to the problem of steady state bifurcations governed by a single nonlinear differential equation of the form

$$(33) \qquad\qquad \frac{d^2y}{dx^2} + F(y,\frac{dy}{dx},\alpha) = 0$$

where the linear differential operator is required to be of the form

$$(34) \qquad\qquad L = \frac{d^2}{dx^2} + constant$$

We assume that the differential equation is defined on a one-dimensional space interval with either zero or zero-flux boundary conditions. Note that if we answer "N" to the question of whether some Taylor coefficients are known to be zero a priori, the program runs fully automatically. However we may reduce the execution time dramatically by setting to zero those coefficients of g and w which vanish for symmetry reasons. We assume that $y \equiv 0$ is a trivial solution of the differential equation for all α, and that the bifurcation parameter α enters the problem only linearly such that no terms of the form $AMP^I LAM^J$ for $J > 1$ exist in g and w.

Here is a sample run on the Euler buckling problem. Since this problem
is invariant when y is changed to – y, all odd order derivatives with respect
to amplitude A (= AMP) in w and g are zero:

```
REDUCTION1();

ENTER DEPENDENT VARIABLE
Y;

USE X AS THE INDEPENDENT VARIABLE AND ALPHA AS A PARAMETER TO VARY
ENTER THE CRITICAL BIFURCATION VALUE ALPHA
%PI^2;
```

$$WE\ DEFINE\ LAM = ALPHA - \%PI^2$$

```
ENTER THE CRITICAL EIGENFUNCTION
COS(%PI*X);
WHAT IS THE LENGTH OF THE X-INTERVAL
1;

SPECIFY THE BOUNDARY CONDITIONS
YOUR CHOICE FOR THE B.C. ON Y AT X=0 AND X= 1
ENTER 1 FOR Y=0, 2 FOR Y'=0
B.C. AT 0?
2;
B.C. AT 1 ?
2;
THE D.E. IS OF THE FORM Y'' + F(Y,Y',ALPHA) = 0, ENTER F
ALPHA*SIN(Y);
```

$$\frac{d^2 Y}{dX^2} + (LAM + \%PI^2)\ SIN(Y)$$

```
DO YOU KNOW A PRIORI THAT SOME TAYLOR COEFFICIENTS ARE ZERO, Y/N
Y;
TO WHICH ORDER DO YOU WANT TO CALCULATE
5;
IS DIFF(W,AMP, 2, LAM, 0) IDENTICALLY ZERO, Y/N
Y;
IS DIFF(W,AMP, 3, LAM, 0) IDENTICALLY ZERO, Y/N
N;
```

$$\left[\frac{d^3 W}{dAMP^3} = - \frac{COS(3\ \%PI\ X)}{32} \right]$$

Compare this result to eqs.(9) and (31).

IS DIFF(W,AMP, 4, LAM, 0) IDENTICALLY ZERO, Y/N
Y;
IS DIFF(W,AMP, 1, LAM, 1) IDENTICALLY ZERO, Y/N
N;

$$\left[\frac{d^2 W}{dAMP\, dLAM} = 0\right]$$

IS DIFF(W,AMP, 2, LAM, 1) IDENTICALLY ZERO, Y/N
Y;
IS DIFF(W,AMP, 3 LAM, 1 IDENTICALLY ZERO, Y/N
N;

$$\left[\frac{d^4 W}{dAMP^3\, dLAM} = -\frac{9\,COS(3\,\%PI\,X)}{256\,\%PI^2}\right]$$

IS G_POLY(1,0) IDENTICALLY ZERO, Y/N
Y;
IS G_POLY(2,0) IDENTICALLY ZERO, Y/N
Y;
IS G_POLY(3,0) IDENTICALLY ZERO, Y/N
N;
IS G_POLY(4,0) IDENTICALLY ZERO, Y/N
Y;
IS G_POLY(5,0) IDENTICALLY ZERO, Y/N
N;
IS G_POLY(1,1) IDENTICALLY ZERO, Y/N
N;
IS G_POLY(2,1) IDENTICALLY ZERO, Y/N
Y;
IS G_POLY(3,1) IDENTICALLY ZERO, Y/N
N;
IS G_POLY(4,1) IDENTICALLY ZERO, Y/N
Y;

$$-\frac{AMP^3\,LAM}{16} + \frac{AMP\,LAM}{2} + \frac{3\,\%PI^2\,AMP^5}{1024} - \frac{\%PI^2\,AMP^3}{16}$$

[VAX 8500 TIME = 390 SEC.]

*This is the bifurcation equation. We manipulate this result to be able to
compare with our introductory calculations:*

SOLVE(%,LAM);

$$\left[LAM = \frac{3\,\%PI^2\,AMP^4 - 64\,\%PI^2\,AMP^2}{64\,AMP^2 - 512}\right]$$

TAYLOR(%,AMP,0,4);

$$
/T/ \qquad [LAM + \ldots = \frac{\%PI^2 \; AMP^2}{8} + \frac{(5 \; \%PI^2) \; AMP^4}{512} + \ldots]
$$

This result agrees with equation (11).

Here is a listing of the MACSYMA functions:

```
/* This file contains REDUCTION1(), a function to perform a Liapunov-Schmidt
reduction for steady state bifurcations of nonlinear d.e.'s of the form
Y'' + F(Y,Y',ALPHA) = 0.  Y = Y(X) is defined on a real interval with Dirichlet
or Neumann boundary conditions and F depends only linearly on ALPHA.
It also contains these additional functions:
SETUP() allows the problem to be entered.
G_POLY(I,J) calculates the coefficient of AMP^I LAM^J in the bifurcation
equation.
W_POLY(I,J) calculates the coefficient of AMP^I LAM^J in W(X;AMP,LAM).
PROJECT(EXP) ensures that <CFUN,EXP>=0.
NEUMANNBC(EXP) accounts for Neumann boundary conditions.
G_EQ() assembles the bifurcation equation. */


REDUCTION1():=BLOCK(

        SETUP(),

        ORDER:READ("TO WHICH ORDER DO YOU WANT TO CALCULATE"),

        FOR I:2 THRU ORDER-1 DO  W_POLY(I,0),

        FOR I:1 THRU ORDER-2 DO  W_POLY(I,1),

        FOR I:1 THRU ORDER DO G_POLY(I,0),

        FOR I:1 THRU ORDER-1 DO G_POLY(I,1),

        G_EQ())$
```

```
SETUP():=(

/* The function SETUP asks for the variables of the d.e., the bifurcation

point, the critical eigenfunction, the x-interval, the boundary conditions and

the differential equation. */

ASSUME_POS:TRUE,

LS_LIST:[],

Y:READ("ENTER DEPENDENT VARIABLE"),

PRINT("USE X AS THE INDEPENDENT VARIABLE AND ALPHA AS A PARAMETER TO VARY"),

CAL:READ("ENTER THE CRITICAL BIFURCATION VALUE ALPHA"),

PRINT("WE DEFINE LAM = ALPHA - ",CAL),

CFUN: READ("ENTER THE CRITICAL EIGENFUNCTION"),

DEPENDS([ZW,Y,W],[AMP,LAM,X]),

LEN:READ("WHAT IS THE LENGTH OF THE X-INTERVAL"),

NORM:INTEGRATE(CFUN^2,X,0,LEN),

PRINT("SPECIFY THE BOUNDARY CONDITIONS"),

PRINT("YOUR CHOICE FOR THE B.C. ON Y AT X=0 AND X=",LEN),

PRINT("ENTER 1 FOR Y=0, 2 FOR Y'=0"),

BCL:READ("B.C. AT 0?"),

BCR:READ("B.C. AT",LEN,"?"),

EQ:DIFF(Y,X,2)

   + READ("THE D.E. IS OF THE FORM Y'' + F(Y,Y',ALPHA) = 0, ENTER F"),

EQLAM:EV(EQ,ALPHA=LAM+CAL),

PRINT(EQLAM),

DATABASE:[DIFF(W,AMP)=0,DIFF(W,LAM)=0],

SUB:Y=AMP*CFUN+W,

TEMP1:EV(EQLAM,SUB,DIFF),

NULLANS:

   READ("DO YOU KNOW A PRIORI THAT SOME TAYLOR COEFFICIENTS ARE ZERO, Y/N") )$
```

```
G_POLY(I,J):=BLOCK(
```

/* This is a function to determine a particular Taylor coefficient of the bifurcation equation G(AMP,LAM) =0. It requires knowledge about the Taylor coefficients of W(AMP,LAM). This knowledge is stored in the list DATABASE. */

```
        LS_LIST:CONS([K=I,L=J],LS_LIST),

        IF NULLANS = Y THEN (

                ZEROANS:READ("IS G_POLY(",I,",",J,") IDENTICALLY ZERO, Y/N"),

                IF ZEROANS = Y THEN  RETURN(BIFEQ[I,J]:0)),

        TEMP2:DIFF(TEMP1,AMP,I,LAM,J),

        DERIVSUBST:TRUE,
```

/* This derivative of w will be annihilated by the projection onto the critical eigenfunction, hence we set it to zero here. */

```
        TEMP3:SUBST('DIFF(W,AMP,I,LAM,J)=0,TEMP2),
```

/* Here we insert all knowledge gained through W_POLY */

```
        D_BASE_LENGTH:LENGTH(DATABASE),

        FOR II THRU D_BASE_LENGTH DO

                TEMP3:EV(SUBS(DATABASE[D_BASE_LENGTH+1-II],TEMP3),DIFF),

        DERIVSUBST:FALSE,

        TEMP4:EV(TEMP3,AMP=0,LAM=0,W=0),
```

/* Projection onto CFUN, the critical eigenfunction. */

```
        TEMP5:TRIGREDUCE(CFUN*TEMP4),

        BIFEQ[I,J]:RATSIMP(INTEGRATE(TEMP5,X,0,LEN))

        )$
```


```
W_POLY(I,J):=BLOCK(
```

/* This function allows the iterative determination of any particular Taylor coefficient of the function W(AMP,LAM). The result is fed into DATABASE. */

```
  IF NULLANS = Y THEN (

        ZEROANS:READ("IS DIFF(W,AMP,",I,", LAM,",J,") IDENTICALLY ZERO ,Y/N"),
```

```
      IF ZEROANS = Y THEN

              (ADDBASE:['DIFF(W,AMP,I,LAM,J)=0],

                      DATABASE:APPEND(DATABASE,ADDBASE),

                      RETURN(ADDBASE))),

   TEMP2:DIFF(TEMP1,AMP,I,LAM,J),

   DERIVSUBST:TRUE,
```

/* We substitute ZW for the unknown Taylor coefficient and solve an o.d.e. for
ZW in TEMP7 */

```
              TEMP3:SUBST(DIFF(W,AMP,I,LAM,J)=ZW,TEMP2),
```

/* Now we insert all knowledge gained through previous iterations. */

```
              D_BASE_LENGTH:LENGTH(DATABASE),

              FOR II THRU D_BASE_LENGTH DO

                      TEMP3:EV(SUBST(DATABASE[D_BASE_LENGTH+1-II],TEMP3),DIFF),

              DERIVSUBST:FALSE,

              TEMP4:EV(TEMP3,AMP=0,LAM=0,W=0),

              TEMP5:TRIGREDUCE(TEMP4),
```

/* This ensures that the r.h.s. of the d.e. TEMP6 is orthogonal to the solution
of the homogeneous equation. */

```
              TEMP6:PROJECT(TEMP5),

              TEMP7:ODE2(TEMP6,ZW,X),
```

/* Satisfy boundary conditions */

```
              IF BCL*BCR=1 THEN TEMP8:BC2(TEMP7,X=0,ZW=0,X=LEN,ZW=0)

                      ELSE

                      TEMP8:NEUMANNBC(TEMP7),
```

/* Make sure that <CFUN,W>=0 */

```
              TEMP9:PROJECT(TEMP8),

              ADDBASE:['DIFF(W,AMP,I,LAM,J)=RHS(TEMP9)],

              DATABASE:APPEND(DATABASE,ADDBASE),

              PRINT(ADDBASE))$
```

```
PROJECT(EXP):=(

        PRO1:EV(EXP,ZW=0),

        PRO2:INTEGRATE(PRO1*CFUN,X,0,LEN)/NORM,

        EXPAND(EXP-PRO2*CFUN))$


NEUMANNBC(EXP):=(

        REXP:RHS(EXP),

        NBC1:DIFF(REXP,X),

        IF BCL=1 AND BCR=2 THEN

                NBC2:SOLVE([EV(REXP,X=0),EV(NBC1,X=LEN)],[%K1,%K2]),

        IF BCL=2 AND BCR=1 THEN

                NBC2:SOLVE([EV(REXP,X=LEN),EV(NBC1,X=0)],[%K1,%K2]),

        IF BCL=2 AND BCR=2 THEN

                NBC2:SOLVE([EV(NBC1,X=LEN),EV(NBC1,X=0)],[%K1,%K2]),

        EV(EXP,NBC2))$


G_EQ():=

        SUM(EV(AMP^K*LAM^L/K!*BIFEQ[K,L],LS_LIST[N]),N,1,LENGTH(LS_LIST))$
```

The following theoretical results have gone into the design of the functions G_POLY(I,J) and W_POLY(I,J):

(i) We want to study the bifurcation from $y = 0$ at $\lambda = 0$. With $y(A,\lambda) = A y_1 + w(x;A,\lambda)$ and $y(0,0) = 0$ we have that $w(x;0,0) = 0$.

(ii) From (16) we see that $Q \tilde{F}(A y_1 + w, \lambda)$ does not contain a simple linear term in $A y_1$. Hence $L w_A = 0$ and since L is invertible on R, $w_A = 0$. This is the first entry into the list DATABASE. Since we assume that $y \equiv 0$ is a trivial solution, we have that $F(0,0,\alpha) = 0$ and hence from differentiating (16.2) with respect to λ and evaluating at $\lambda = A = 0$, we find that $w_\lambda \equiv 0$. This is the second entry into DATABASE.

(iii) Eq(16.1) does not contain the term L w since P L w = 0. This is

obtained in G_POLY() at the stage TEMP3 where we set $\dfrac{d^{i+j}w}{dA^i d\lambda^j} = 0$, since this

term can only occur linearly (cf. the argument leading to Eq.(23)).

(iv) The same term $\dfrac{d^{i+j}w}{dA^i d\lambda^j}$ is the unknown variable in W_POLY for the

Taylor coefficient $A^i \lambda^j$ in $w(A,\lambda)$. We replace it by the variables ZW in TEMP3
of W_POLY().

(v) The projection Q is performed by the function PROJECT(EXP). Since
the linear problem is self-adjoint, the projection is given by

$$Q(exp) = exp - \frac{\int_0^{LEN} exp \; CFUN \; dx}{NORM} \; CFUN$$

with NORM = $\int_0^{LEN} CFUN^2 \; dx$. This ensures that the ordinary differential equation

TEMP6 has a nontrivial particular solution. The ode-solver in MACSYMA, ODE2,
adds the solution of the homogeneous equation to any particular solution it
finds. After having fitted the boundary conditions we then have to perform the
Q-projection again. This ensures that the projection of w onto the critical
eigenfunction <CFUN,W> is always zero. For the Euler buckling problem (indeed
for all problems with Neumann boundary conditions at both ends of the defining
interval), this final projection is trivial and amounts to deleting the term
which contains the critical eigenfunction explicitly. However, in some cases
(e.g. quadratic nonlinearities with Dirichlet boundary conditions) we encounter
resonance terms which make the projection Q nontrivial (cf. Exercise 4).

Liapunov-Schmidt Reduction for O.D.E's

In the following Sections we will increase the complexity of the
Liapunov–Schmidt reduction program in steps. While we dealt in the last
Section with bifurcations of steady–state solutions in one partial differential
equation depending on one space variable, our next step will be to treat n
coupled differential equations. As a starting point we confine ourselves to n
coupled ordinary differential equations. Since we consider only steady–state
solutions, we are effectively dealing with purely algebraic equations. Hence a
Liapunov–Schmidt reduction for a steady state bifurcation in a system of
o.d.e.'s is very simple. The projections P and Q are simple products of two
$\mathbb{R}^n$–vectors involving the eigenvector corresponding to the eigenvalue zero.
Inverting the linear operator L amounts to inverting a matrix or solving a
linear system of algebraic equations.

Why would one want to perform a Liapunov–Schmidt reduction on ordinary
differential equations when techniques are available (notably the center
manifold reduction) which allow not only the determination of the bifurcated
steady state solutions but also the dynamical behavior of the solutions
approaching these steady states. We offer three answers to this question:
First, it is instructive to see to what degree both methods give the same
results and when and how they differ. Second, there are criteria (see below)
which tell when the result of a Liapunov–Schmidt reduction can be augmented
with some simple dynamics. And third, a Liapunov–Schmidt reduction often
involves less computation than a center manifold reduction.

In order to explain the latter fact, let us imagine a steady state
bifurcation in an n–dimensional dynamical system (n say > 5). If the
bifurcation is nondegenerate, then the center manifold is one–dimensional. In
order to calculate the flow on the center manifold we perform nonlinear
coordinate transformations on the whole n–dimensional system, although we are

finally interested only in one Taylor coefficient, say the third order term for the flow on the center manifold. Since the Liapunov–Schmidt reduction permits us to find a specific coefficient in the bifurcation equation directly, less computation time is required for the Liapunov–Schmidt reduction than for the center manifold reduction. This is supported by experience with both methods for nontrivial research problems. Problems where a MACSYMA program for a center manifold reduction could not finish in a reasonable time were able to be handled by a Liapunov–Schmidt reduction program.

To show the equivalence and differences of both methods we consider again the Lorenz system of Chapter 2.

$$x_1' = \sigma \, (x_2 - x_1)$$

(35)
$$x_2' = \rho \, x_1 - x_2 - x_1 x_3$$

$$x_3' = - \beta \, x_3 + x_1 x_2$$

We calculated in Chapter 2 that for $\rho = 1$ one of the eigenvalues of (35) is zero with the corresponding critical eigenvector $[1,1,0]$. The linear operator L at $\rho = \rho_c = 1$ associated with (35) is the matrix:

$$L = \begin{bmatrix} -\sigma & \sigma & 0 \\ 1 & -1 & 0 \\ 0 & 0 & -\beta \end{bmatrix}$$

$$\text{and } L^* = L^t = \begin{bmatrix} -\sigma & 1 & 0 \\ \sigma & -1 & 0 \\ 0 & 0 & -\beta \end{bmatrix},$$

is the adjoint matrix. This leads to the adjoint critical eigenfunction $[1/\sigma, 1, 0]$. The following is a sample run of a MACSYMA program REDUCTION2 to

determine the bifurcation equation for this instability. See Exercise 1 at the
end of the Chapter.

REDUCTION2( );

NUMBER OF EQUATIONS
3;
ENTER VARIABLE NUMBER 1
X1;
ENTER VARIABLE NUMBER 2
X2;
ENTER VARIABLE NUMBER 3
X3;
ENTER THE BIFURCATION PARAMETER
RHO;
ENTER THE CRITICAL BIFURCATION VALUE RHO
1;

WE DEFINE LAM = RHO - 1
ENTER THE CRITICAL EIGENVECTOR AS A LIST
[1,1,0];
ENTER THE ADJOINT CRITICAL EIGENVECTOR
[1/SIGMA,1,0];

ENTER THE DIFFERENTIAL EQUATION
DIFF(X1 ,T)=
SIGMA*(X2-X1);
DIFF(X2 ,T)=
X1*X3+RHO*X1-X2;
DIFF(X3 ,T)=
X1*X2-BETA*X3;

[SIGMA (X2 - X1),- X1 X3 - X2 + (LAM + 1) X1, X1 X2 - BETA X3]
DO YOU KNOW A PRIORI THAT SOME TAYLOR COEFFICENTS ARE ZERO, Y/N
N;
TO WHICH ORDER DO YOU WANT TO CALCULATE
3;

$$
\left[\frac{d^2 W1}{dAMP^2} = 0, \quad \frac{d^2 W2}{dAMP^2} = 0, \quad \frac{d^2 W3}{dAMP^2} = \frac{2}{BETA}\right]
$$

$$
\left[\frac{d^2 W1}{dAMP\ dLAM} = -\frac{SIGMA}{SIGMA^2 + 2\ SIGMA + 1}, \quad \frac{d^2 W2}{dAMP\ dLAM} = \frac{1}{SIGMA^2 + 2\ SIGMA + 1},\right.
$$

$$
\left.\frac{d^2 W3}{dAMP\ dLAM} = 0\right]
$$

[SYMBOLICS 3670 TIME = 95 SEC.]

$$\text{AMP LAM} - \frac{\text{AMP}^3}{\text{BETA}}$$

We find a bifurcation equation

(36)
$$g_{\text{Lorenz}}(A,\lambda) = \lambda\, A - (1/\beta)\, A^3$$

with nontrivial solutions $A = \pm\,(\beta\,\lambda)^{(1/2)}$ which are identical to the solutions found for the steady states on the center manifold (eq.(28) in Chapter 2). In general these two reductions do not lead to identical results for the steady state solutions but the topological character of the branching (i.e. here a pitchfork bifurcation) is the same in both cases. However one can show (cf.[13]) that if $\langle y_1, y_1^* \rangle > 0$ then the stability of the solutions of the bifurcation equation $g(A,\lambda) = 0$ with respect to perturbations in the direction of the critical eigenfunction are given by the stability of the corresponding solutions of the dynamical system

(37)
$$\frac{dA}{dt} = g(A,\lambda)$$

Non-uniqueness of the Liapunov–Schmidt reduction and a change of stability is demonstrated by choosing a different critical eigenvector. We multiply the critical eigenvector by a term (–const) and perform the reduction:

REDUCTION2();

some input lines omitted for brevity ...

ENTER THE CRITICAL EIGENVECTOR AS A LIST
[-CONST,-CONST,0];

some input lines omitted for brevity ...

[SIGMA (X2 - X1),- X1 X3 - X2 + (LAM + 1) X1, X1 X2 - BETA X3]
DO YOU KNOW A PRIORI THAT SOME TAYLOR COEFFICENTS ARE ZERO, Y/N
N;

TO WHICH ORDER DO YOU WANT TO CALCULATE
3;

$$\left[\frac{d^2 W1}{dAMP^2} = 0, \quad \frac{d^2 W2}{dAMP^2} = 0, \quad \frac{d^2 W3}{dAMP^2} = \frac{2\ CONST}{BETA}\right]$$

$$\left[\frac{d^2 W1}{dAMP\ dLAM} = -\frac{CONST\ SIGMA}{SIGMA^2 + 2\ SIGMA + 1}, \quad \frac{d^2 W2}{dAMP\ dLAM} = \frac{CONST}{SIGMA^2 + 2\ SIGMA + 1},\right.$$

$$\left.\frac{d^2 W3}{dAMP\ dLAM} = 0\right]$$

[SYMBOLICS 3670 TIME = 72 SEC.]

$$- AMP\ CONST\ LAM + \frac{AMP^3\ CONST^3}{BETA}$$

We find

(38) $\tilde{g}_{Lorenz}(A,\lambda) = -\ const\ \lambda\ A + (const^3/\beta)\ A^3 = 0$

as the bifurcation equation which has nontrivial solutions $A = \pm \dfrac{(\lambda\ \beta)^{(1/2)}}{const}$.

Let us determine the stability of the trivial solution $A = 0$ for the two reduced systems $\dfrac{dA}{dt} = g_{Lorenz}$, $\dfrac{dA}{dt} = \tilde{g}_{Lorenz}$ and of the corresponding trivial solution $x_1 = x_2 = x_3 = 0$ of the original Lorenz system (35). For $\rho \to -\infty$ the eigenvalues of these systems become $-\infty$, sign(const) ∞, and $\left[-\dfrac{\sigma + 1}{2} - i\ \infty, -\dfrac{\sigma + 1}{2} + i\ \infty, -\beta\right]$ (cf. Chapter 2), respectively. Hence for $\sigma > -1$ and const > 0 we confirm that the solutions of $\dfrac{dA}{dt} = g_{Lorenz}$ have the same stability as the corresponding solutions of the original system of o.d.e.'s. Also since the norm $\langle y_1^*, y_1 \rangle = 1 + 1/\sigma$, we find that if $\sigma < -1$ and const > 0 then the solutions of $\dfrac{dA}{dt} = \tilde{g}_{Lorenz}$ have the same stability as the corresponding solutions of the Lorenz system, thus verifying the aforementioned proposition of [13].

Application to Systems of Partial Differential Equations

Clearly there is no hope of writing a general program in MACSYMA which can deal with bifurcations in all physically interesting partial differential equations (p.d.e.'s) which may range over such fields as general relativity, plasmaphysics, hydrodynamics, elasticity and reaction diffusion equations. The major restriction to that task is not lack of memory space or slow performance (although this may become important), but rather the absence of suitable abstract datatypes in MACSYMA. E.g. it is difficult to write a Frechet derivative and it is not clear how to define function spaces like a Banach space, etc. Hopefully the next generation of computer algebra systems (whose first member is SCRATCHPAD II [46]) will give us the tools to work with differential operators on specifically defined function spaces.

Therefore the program which we are going to describe now is designed to work for steady state bifurcations of a system of p.d.e.'s of the form

$$\frac{\partial}{\partial t} = \text{Lin}\left[\frac{\partial^2}{\partial x^2}\right] + \text{NL}$$

where $\text{Lin}\left[\frac{\partial^2}{\partial x^2}\right]$ is a linear differential operator of even order defined on some function space of $\mathbb{R}$, and NL is a nonlinear (differential) operator. A typical example of the type of equations which can be handled by this program are coupled reaction diffusion equations depending on one spatial dimension. Since the differential equation solver ODE2 in MACSYMA works only for one-dimensional o.d.e's of first or second order, we solve the differential equation for the Taylor coefficients of $w(x;a,\lambda)$, eq.(16.2), via a Fourier mode ansatz. However this has its own problems: Partial differential equations with Dirichlet boundary conditions and quadratic nonlinearities lead to ordinary differential equations for the Taylor coefficient $\dfrac{d^2w}{dA^2}$ which have no finite Fourier mode

expansion satisfying the boundary conditions (see Exercise 4 at the end of the

Chapter.) Instead, these o.d.e's display "resonant" terms. Solutions to these

equations can readily be found for a single ordinary differential equation (see

Exercise 4). For n-coupled o.d.e.'s, this is not so easy and certainly goes

beyond the scope of this book. Therefore the program we are giving here

cannot, in general, handle quadratic nonlinearities with sine-functions as

critical eigenfunctions of the linearized problem. (See however the specific

case in Exercise 5).

The following example taken from Carr [6] shall demonstrate the

program. Let us consider the semilinear wave equation

$$(39) \qquad v_{tt} + v_t - v_{xx} - \alpha\, v + f(v) = 0, \qquad f(v) = v^3 + \rho\, v^5,$$

with $v = 0$ at $x = 0, \pi$. Eq.(39) can be rewritten setting $y_1 = v$, $y_2 = v_t$ as

two coupled first order equations,

$$(40) \qquad \begin{aligned} y_{1_t} &= y_2 \\[2mm] y_{2_t} &= y_{1_{xx}} + \alpha\, y_1 - y_2 - y_1^{\,3} - \rho\, y_1^{\,5} \end{aligned}$$

The linear operator $L(\alpha)$ is given by

$$(41) \qquad L(\alpha) = \begin{bmatrix} 0 & 1 \\ d_{xx} + \alpha & -1 \end{bmatrix}$$

where d_{xx} denotes the second partial derivative with respect to the spatial

variable. The eigenvalues of the linearized system are

$$EV_n^{+-} = \frac{1}{2} \left[-1 \overset{+}{-} \sqrt{1 - 4 \, (n^2 - \alpha)} \, \right]$$

for eigenfunctions of the form $\begin{bmatrix} EF_1 \\ EF_2 \end{bmatrix}$ sin nx. Therefore for the interval

$0 \leq \alpha < 1$, all eigenvalues have negative real parts and the trivial solution is stable. For $\alpha = 1$ the eigenvalue EV_1^+ is zero and the n = 1 mode becomes unstable. The corresponding eigenfunction is CFUN = $\begin{bmatrix} 1 \\ 0 \end{bmatrix}$ sin x. Setting

$u = \begin{bmatrix} u_1 \\ u_2 \end{bmatrix}$ and $v = \begin{bmatrix} v_1 \\ v_2 \end{bmatrix}$ we find

$$\langle u, L(\alpha) \, v \rangle = \int_0^\pi u_1 v_2 + u_2(\alpha v_1 + v_{1_{xx}} - v_2) \; dx$$

$$= \int_0^\pi u_1 v_2 + u_2(\alpha v_1 - v_2) + u_{2_{xx}} v_1 \; dx \qquad \text{(integration by parts)}$$

$$= \int_0^\pi v_1(\alpha u_2 + u_{2_{xx}}) + v_2(u_1 - u_2) \; dx$$

$$= \langle L^*(\alpha) \, u, v \rangle$$

where we define the adjoint linear operator $L^*(\alpha)$ by

$$L^*(\alpha) = \begin{bmatrix} 0 & d_{xx} + \alpha \\ 1 & -1 \end{bmatrix}$$

Hence the adjoint critical eigenfunction is given by $y_1^* = A \sin x \begin{bmatrix} 1 \\ 1 \end{bmatrix}$. We normalize such that $\langle y_1^*, y_1 \rangle = 1$ which gives us AFUN $= \frac{2}{\pi} \sin x \begin{bmatrix} 1 \\ 1 \end{bmatrix}$.

To calculate the bifurcation equation for (40) near $\alpha = 1$ we make the

now familiar ansatz $y = \begin{bmatrix} y_1 \\ y_2 \end{bmatrix} = A \sin(x) \begin{bmatrix} 1 \\ 0 \end{bmatrix} + \begin{bmatrix} w_1(x;A,\lambda) \\ w_2(x;A,\lambda) \end{bmatrix}$ with $\alpha = \lambda + 1$ and

substitute this into (40).

Let us demonstrate how we can calculate the two Taylor coefficients $\dfrac{d^2w}{dA^2}$

and $\dfrac{d^3w}{dA^3}$ as an example. We show that the former coefficient is always zero: By

entering the ansatz into (40), differentiating twice with respect to A and

setting $A = \lambda = 0$, we obtain the two equations

$$0 = \frac{d^2w_2}{dA^2}\bigg|_{\substack{A=0\\\lambda=0}}$$

$$0 = \left[\frac{d^2w_1}{dA^2} - \frac{d^2w_2}{dA^2} + \frac{d^2w_1}{dA^2dx^2}\right]\Bigg|_{\substack{A=0\\\lambda=0}}$$

which may be written in the form

$$0 = L(1)\,\frac{d^2}{dA^2}\begin{bmatrix}w_1\\w_2\end{bmatrix}\Bigg|_{\substack{A=0\\\lambda=0}}$$

Now since $L(1)\,y = 0$ has only y_1 as nontrivial solution and since we require

that $\langle y_1^*, w\rangle$ is zero, we have that the Taylor coefficient $\dfrac{d^2w}{dA^2}\bigg|_{\substack{A=0\\\lambda=0}}$ is zero.

We could have concluded this right from the beginning, since (40) is invariant

with respect to a reflection $y \to -y$ and hence the bifurcation equation and the

function $w(x;A,\lambda)$ inherit this symmetry and are equivariant with respect to the

transformation $A \to -A$ [43], i.e. all terms of even order in A are zero.

Our next step is to find the third order coefficient of w with respect

to A. If we set $zw_1 = \dfrac{d^3w_1}{dA^3}\bigg|_{\substack{A=0\\\lambda=0}}$ and $zw_2 = \dfrac{d^3w_2}{dA^3}\bigg|_{\substack{A=0\\\lambda=0}}$, we find after elementary

calculations the following two equations for zw_1 and zw_2:

(42.1) $zw_2 = 0$

(42.2) $\dfrac{d^2 zw_1}{dx^2} - zw_2 + zw_1 = 9/2 \sin x - 3/2 \sin 3x$

Substituting (42.1) into (42.2) and using the boundary conditions $zw_1 = 0$ at

$x = 0, \pi$, we see that (42.2) has no solution. By the Fredholm alternative

theorem, since the homogeneous equation has a nontrivial solution, the

nonhomogeneous equation will have a solution only if the nonhomogenity is

orthogonal to the null space of the adjoint operator $L^*(1)$. This is guaranteed

(cf. eq.(14.2)) by replacing $\begin{bmatrix} 0 \\ 9/2 \sin x - 3/2 \sin 3x \end{bmatrix}$ by

(43) $\begin{bmatrix} 0 \\ 9/2 \sin x - 3/2 \sin 3x \end{bmatrix} - \dfrac{\left\langle \begin{bmatrix} 0 \\ 9/2 \sin x - 3/2 \sin 3x \end{bmatrix}, y_1^* \right\rangle y_1}{\langle y_1, y_1^* \rangle}$

Then $\langle y_1^*, \text{eq.}(43)\rangle$ is always zero. We calculate

$$2/\pi \int_0^\pi \{9/2 \sin^2 x - 3/2 \sin 3x \sin x\}\, dx = 9/2.$$

Hence the Q–projected differential equations (42) become

$$zw_2 = - 9/2 \sin x$$

$$\dfrac{d^2 zw_1}{dx^2} + zw_1 - zw_2 = 9/2 \sin x - 3/2 \sin 3x$$

and a general solution to these equations with the zero boundary conditions at

$x = 0, \pi$ is given by

$$zw_1 = k_1 \sin x + 3/16 \sin 3x$$

$$zw_2 = -9/2 \sin x$$

where k_1 is an undetermined constant.

Now since we require $\langle y_1{}^*, w \rangle = 0$, we have to perform the same projection on this solution again, giving us, with

$$\langle y_1{}^*, \begin{bmatrix} k_1 \sin x + 3/16 \sin 3x \\ -9/2 \sin x \end{bmatrix} \rangle = -9/2 + k_1$$

the third order Taylor coefficient

$$\begin{bmatrix} \dfrac{d^3 w_1}{dA^3} \Big|_{\substack{A=0 \\ \lambda=0}} \\ \dfrac{d^3 w_2}{dA^3} \Big|_{\substack{A=0 \\ \lambda=0}} \end{bmatrix} = \begin{bmatrix} 3/16 \sin 3x + 9/2 \sin x \\ \\ -9/2 \sin x \end{bmatrix}$$

Since the fourth order term is again zero for symmetry reasons, we can now calculate the Taylor coefficients of A for the bifurcation equation $g(A,\lambda) = 0$ up to fifth order. Evaluating the fifth order term is in principle the same as for the one-dimensional bifurcation problems treated previously. However the projection P of eq.(14.1) now involves an inner product (13) which includes the scalar product with the adjoint eigenfunction y_1^* in $\mathbb{R}^2$.

At this point we shall present a MACSYMA program, REDUCTION3, which accomplishes the Liapunov–Schmidt reduction for n coupled d.e.'s in one spatial variable. We begin with a sample run on the bifurcation problem of eq.(40), followed by the program listing:

REDUCTION3();

ENTER THE NUMBER OF DIFFERENTIAL EQUATIONS
2;
ENTER THE DEPENDENT VARIABLES AS A LIST
[Y1,Y2];
ENTER THE SPATIAL COORDINATE
X;
ENTER THE BIFURCATION PARAMETER
ALPHA;
ENTER THE CRITICAL BIFURCATION VALUE
1;
WE DEFINE LAM = ALPHA - 1
ENTER THE CRITICAL EIGENFUNCTION AS A LIST
SIN(X)*[1,0];
ENTER THE ADJOINT CRITICAL EIGENFUNCTION AS A LIST
2/%PI*SIN(X)*[1,1];
ENTER THE DIFFERENTIAL EQUATION NUMBER 1
Y2;
ENTER THE DIFFERENTIAL EQUATION NUMBER 2
DIFF(Y1,X,2)+ALPHA*Y1-Y2-Y1^3-A*Y1^5;

$$\left[Y2, \ -\ Y2\ +\ \frac{d^2 Y1}{dX^2}\ -\ A\ Y1^5\ -\ Y1^3\ +\ (LAM\ +\ 1)\ Y1\right]$$

WHAT IS THE LENGTH OF THE SPACE INTERVAL
%PI;
DO YOU KNOW A PRIORI THAT SOME TAYLOR COEFFICIENTS ARE 0, Y/N
Y;
TO WHICH ORDER DO YOU WANT TO CALCULATE
5;
IS DIFF(W(AMP, 2 ,LAM, 0) IDENTICALLY ZERO, Y/N
Y;
IS DIFF(W(AMP, 3 ,LAM, 0) IDENTICALLY ZERO, Y/N
N;

$$\left[\frac{d^3 W1}{dAMP^3} = \frac{3\ SIN(3\ X)\ +\ 72\ SIN(X)}{16},\ \ \frac{d^3 W2}{dAMP^3} = -\ \frac{9\ SIN(X)}{2}\right]$$

IS DIFF(W(AMP, 4 ,LAM, 0) IDENTICALLY ZERO, Y/N
Y;
IS DIFF(W(AMP, 1 ,LAM, 1) IDENTICALLY ZERO, Y/N
N;

$$\left[\frac{d^2 W1}{dAMP\ dLAM} = -\ SIN(X),\ \ \frac{d^2 W2}{dAMP\ dLAM} = SIN(X)\right]$$

IS DIFF(W(AMP, 2 ,LAM, 1) IDENTICALLY ZERO, Y/N
Y;

IS DIFF(W(AMP, 3 ,LAM, 1) IDENTICALLY ZERO, Y/N
N;

$$\left[\frac{d^4 W1}{dAMP^3 \; dLAM} = -\frac{69 \; SIN(3 \; X) + 2304 \; SIN(X)}{128}, \quad \frac{d^4 W2}{dAMP^3 \; dLAM} = 18 \; SIN(X)\right]$$

IS G_POLY(1 , 0)IDENTICALLY ZERO, Y/N
Y;
IS G_POLY(2 , 0)IDENTICALLY ZERO, Y/N
Y;
IS G_POLY(3 , 0)IDENTICALLY ZERO, Y/N
N;
IS G_POLY(4 , 0)IDENTICALLY ZERO, Y/N
Y;
IS G_POLY(5 , 0)IDENTICALLY ZERO, Y/N
N;
IS G_POLY(1 , 1)IDENTICALLY ZERO, Y/N
N;
IS G_POLY(2 , 1)IDENTICALLY ZERO, Y/N
Y;
IS G_POLY(3 , 1)IDENTICALLY ZERO, Y/N
N;
IS G_POLY(4 , 1)IDENTICALLY ZERO, Y/N
Y;

[SYMBOLICS 3670 TIME = 328 SEC.]

$$3 \; AMP^3 \; LAM + AMP \; LAM + \frac{(-1200 \; \%PI \; A - 3195 \; \%PI) \; AMP^5}{1920 \; \%PI} - \frac{3 \; AMP^3}{4}$$

This is the bifurcation equation. We manipulate this result to obtain λ as a
function of A:

SOLVE(%,LAM);

$$\left[LAM = \frac{(80 \; A + 213) \; AMP^4 + 96 \; AMP^2}{384 \; AMP^2 + 128}\right]$$

TAYLOR(%,AMP,0,4);

/T/ $$\left[LAM + \ldots = \frac{3 \; AMP^2}{4} + \frac{(80 \; A - 75) \; AMP^4}{128} + \ldots\right]$$

The following is a listing of the functions which perform the Liapunov-Schmidt reduction:

```
/* This file contains REDUCTION3(), a function to perform a Liapunov-Schmidt
reduction for steady state bifurcations from n coupled partial differential
equations defined on one spatial dimension.  The following additional functions
are supplied:
  SETUP() allows the problem to be entered.
  G_POLY(I,J) calculates the coefficient of AMP^I LAM^J in the bifurcation
  equation G(AMP,LAM).
  W_POLY(I,J) calculates the coefficient of AMP^I LAM^J in W(X;AMP,LAM).
  SOLVE_ODE(EXP) solves certain ordinary differential equations via a Fourier
  mode ansatz.
  FEEDW(EXP) ensures that <AFUN,W> = 0 .
  FIND_TRIG(EXP) identifies Fourier modes.
  SETIFY(LIST) transforms a list into a set.
  G_EQ() assembles the bifurcation equation.
  VFUN(LIST,VALUE) creates the substitution list:
                    [LIST[1] = VALUE, LIST[2] = VALUE,...]
  DIFFNULL(I,J) sets the derivative diff(w,amp,i,lam,j) to zero. */


REDUCTION3():=BLOCK(
      SETUP(),
      ORDER:READ("TO WHICH ORDER DO YOU WANT TO CALCULATE"),
      FOR I:2 THRU ORDER-1 DO  W_POLY(I,0),
      FOR I:1 THRU ORDER-2 DO  W_POLY(I,1),
      FOR I:1 THRU ORDER DO G_POLY(I,0),
      FOR I:1 THRU ORDER-1 DO G_POLY(I,1),
      G_EQ())$
```

```
SETUP():=(  /* This function performs the input for the Liapunov-Schmidt
reduction*/
ASSUME_POS:TRUE,
LS_LIST:[],
N:READ("ENTER THE NUMBER OF DIFFERENTIAL EQUATIONS"),
Y:READ("ENTER THE DEPENDENT VARIABLES AS A LIST"),
XVAR:READ("ENTER THE SPATIAL COORDINATE"),
ALPHA:READ("ENTER THE BIFURCATION PARAMETER"),
CAL:READ("ENTER THE CRITICAL BIFURCATION VALUE"),
PRINT("WE DEFINE LAM = ",ALPHA - CAL),
CFUN:READ("ENTER THE CRITICAL EIGENFUNCTION AS A LIST"),
AFUN:READ("ENTER THE ADJOINT CRITICAL EIGENFUNCTION AS A LIST"),
KILL(W),
W:MAKELIST(CONCAT(W,I),I,1,N),
ZWLIST:MAKELIST(CONCAT(ZW,I),I,1,N),
NULLIST:MAKELIST(0,I,1,N),
DEPENDS(APPEND(ZWLIST,W,Y),CONS(XVAR,[AMP,LAM])),
EQS:MAKELIST(READ("ENTER THE DIFFERENTIAL EQUATION NUMBER",I),I,1,N),
EQLAM:EV(EQS,EV(ALPHA) = LAM + CAL),
PRINT(EQLAM),
LEN:READ("WHAT IS THE LENGTH OF THE SPACE INTERVAL"),
WNULL:VFUN(W,0),
SUB:MAPLIST("=",Y,AMP*CFUN+W),
DATABASE:APPEND(DIFNULL(1,0),DIFNULL(0,1)),
ZWNULL:VFUN(ZWLIST,0),
NORM:INTEGRATE(AFUN.CFUN,XVAR,0,LEN),
TEMP1:EV(EQLAM,SUB,DIFF),
NULLANS:READ("DO YOU KNOW A PRIORI THAT SOME TAYLOR COEFFICIENTS ARE 0, Y/N")
)$
```

```
G_POLY(I,J):=BLOCK(

/* This is a function to determine a particular Taylor coefficient of the
bifurcation equation G(AMP,LAM) = 0.  It requires kowledge about the Taylor
coefficients of W(AMP,LAM).  This knowledge is stored in the list DATABASE. */

        LS_LIST:CONS([K=I,L=J],LS_LIST),

        IF NULLANS = Y THEN (

                ZEROANS:READ("IS G_POLY(",I,",",J,")IDENTICALLY ZERO, Y/N"),

                IF ZEROANS = Y THEN RETURN(BIFEQ[I,J]:0)),

        TEMP2:DIFF(TEMP1,AMP,I,LAM,J),

        DERIVSUBST:TRUE,

        /* Set the derivatives diff(w,amp,i,lam,j) to zero. */

        TEMP3:SUBST(DIFNULL(I,J),TEMP2),

        /* Enter all information in DATABASE. */

        D_BASE_LENGTH:LENGTH(DATABASE),

        FOR II THRU D_BASE_LENGTH DO

                TEMP3:EV(SUBST(DATABASE[D_BASE_LENGTH+1-II],TEMP3),DIFF),

        DERIVSUBST:FALSE,

        TEMP4:EXPAND(EV(TEMP3,AMP=0,LAM=0,WNULL,INTEGRATE)),

        /* Project onto AFUN. */

        TEMP5:RATSIMP(TEMP4.AFUN),

        BIFEQ[I,J]:INTEGRATE(TRIGREDUCE(TEMP5),XVAR,0,LEN))$


W_POLY(I,J):=BLOCK(

/* This function allows the iterative determination of any particular Taylor
coefficient of the function W(AMP,LAM).  It returns a differential equation for
the particular coefficient of interest (called ZW1,ZW2...).  This d.e. is
solved via SOLVE_ODE and the result is fed into DATABASE from FEEDW. */

        IF NULLANS = Y THEN (

                ZEROANS:READ("IS DIFF(W(AMP,",I,",LAM,",J,") IDENTICALLY ZERO, Y/N"),
```

```
    IF ZEROANS = Y THEN (

                         ADDBASE:DIFNULL(I,J),

                         DATABASE:APPEND(DATABASE,ADDBASE),

                         RETURN(ADDBASE))),

WMAX_DIFF:MAP(LAMBDA([U],'DIFF(U,AMP,I,LAM,J) = CONCAT(Z,U)),W),

TEMP2:DIFF(TEMP1,AMP,I,LAM,J),

DERIVSUBST:TRUE,

/* Rename the derivatives diff(w,amp,i,lam,j) to zw */

TEMP3:SUBST(WMAX_DIFF,TEMP2),

/* Enter all information stored in DATABASE */

D_BASE_LENGTH:LENGTH(DATABASE),

FOR II THRU D_BASE_LENGTH DO

        TEMP3:EV(SUBST(DATABASE[D_BASE_LENGTH+1-II],TEMP3),DIFF),

DERIVSUBST:FALSE,

TEMP4:EV(TEMP3,AMP=0,LAM=0,WNULL,INTEGRATE),

/* This is the projection Q onto the range of the

   linear differential operator in the problem. */

TEMP5:INTEGRATE(EV(TEMP4,ZWNULL).AFUN,XVAR,0,LEN),

TEMP6:TEMP4-TEMP5/NORM*CFUN,

TEMP7:TRIGREDUCE(TEMP6),

/* The set of o.d.e.'s to solve.*/

W_DE:EXPAND(TEMP7),

TEMP8:EV(W_DE,VFUN(ZWLIST,0)),

/* If the particular solution of W_DE is zero then w=0. */

IF NULLIST=TEMP8 THEN ( ADDBASE:DIFNULL(I,J),

                         DATABASE:APPEND(DATABASE,ADDBASE),

                         RETURN(ADDBASE)),

TEMP9:SOLVE_ODE(TEMP8),

FEEDW(TEMP9)    )$
```

```
SOLVE_ODE(EXP):=(/* This function solves the d.e. W_DE by a Fourier mode
ansatz. */
        TRIGFUN:[],
        CONST:FALSE,
        FOR I THRU N DO (
        /* Determine the Fourier modes */
                TRIG1:EXP[I],
                IF TRIG1 # 0 THEN (
                        TRIG2:APPLY1(TRIG1,SINNULL,COSNULL),
                        IF TRIG2 # 0 THEN (
                                CONST:TRUE,
                                TRIG1:TRIG1-TRIG2),
                TRIGFUN:APPEND(FIND_TRIG(TRIG1),TRIGFUN))),
        SOL1:DELETE(DUMMY,SETIFY(TRIGFUN)),
        /* Make an ansatz */
        ANSATZ:MAKELIST(SUM(AM[I,J]*SOL1[I],I,1,LENGTH(SOL1)),J,1,N),
        SOL2:EV(W_DE,MAP("=",ZWLIST,ANSATZ),DIFF),
        SOL3:MAKELIST(RATCOEF(SOL2,I),I,SOL1),
        EQLIST:[],
        FOR I THRU LENGTH(SOL3) DO EQLIST:APPEND(EQLIST,SOL3[I]),
        VARLIST:[],
        FOR I THRU N DO FOR J THRU LENGTH(SOL1) DO
                VARLIST:CONS(AM[J,I],VARLIST),
        /* Find the amplitudes of the Fourier modes */
        SOL4:SOLVE(EQLIST,VARLIST),
        /* Solve for the constant Fourier mode if necessary */
        CANSATZ:0,
        IF CONST = TRUE THEN (CANSATZ:MAKELIST(CONCAT(CON,I),I,1,N),
                        CSOL1:EV(W_DE,MAP("=",ZWLIST,CANSATZ),DIFF),
```

```
                            CSOL2:APPLY1(CSOL1,SINNULL,COSNULL),

                            CSOL3:SOLVE(CSOL2,CANSATZ)),

        EV(ANSATZ+CANSATZ,SOL4,CSOL3))$


FEEDW(EXP):=(/* This function allows the result of the ode-solver to be entered
                into the list DATABASE.  It checks for orthogonality to the
                critical adjoint eigenfunction and removes solutions of the
                homogeneous equation (i.e. nonorthogonal terms). */
        F2:INTEGRATE(EXP.AFUN,XVAR,0,LEN),

        IF RATSIMP(F2)=0

                THEN

                        ADDBASE:MAP("=",MAKELI'DIFF(W[U],AMP,I,LAM,J),U,1,N),EXP)

                        ELSE (ORTHO_RESULT:RATSIMP(EXP- F2/NORM*CFUN),

                                ADDBASE:MAP("=",MAKELIST('DIFF(W[U],AMP,I,LAM,J),U,1,N),

                                        ORTHO_RESULT)),

                DATABASE:APPEND(DATABASE,ADDBASE),

                PRINT(ADDBASE))$
/* Collect all information stored in BIFEQ and assemble the bifurcation
equation. */
G_EQ():=  SUM(EV(AMP^K*LAM^L/K!*BIFEQ[K,L],LS_LIST[N]),N,1,LENGTH(LS_LIST))$


SETIFY(LIST):=(/* Transforms a list into a set, i.e. removes duplicates. */
        SET:[LIST[1]],

        FOR I :2 THRU LENGTH(LIST) DO

                IF NOT MEMBER(LIST[I],SET) THEN

                        SET:CONS(LIST[I],SET),

        SET)$
```

```
FIND_TRIG(EXP):=(/* Finds the Fourier modes. */
        F_A1:ARGS(EXP+DUMMY),
        F_A2:APPLY1(F_A1,SINFIND,COSFIND)
        )$


/* Auxiliary Functions */
MATCHDECLARE([DUMMY1,DUMMY2],TRUE)$
DEFRULE(COSFIND,DUMMY1*COS(DUMMY2),COS(DUMMY2))$
DEFRULE(SINFIND,DUMMY1*SIN(DUMMY2),SIN(DUMMY2))$
DEFRULE(COSNULL,COS(DUMMY1),0)$
DEFRULE(SINNULL,SIN(DUMMY1),0)$


VFUN(LIST,VALUE):=MAP(LAMBDA([U],U=VALUE),LIST)$
DIFNULL(I,J):=MAP(LAMBDA([U],'DIFF(U,AMP,I,LAM,J)=0),W) $
```

Comments on the programs:

The main difference between these programs and those presented earlier in this
Chapter is just bookkeeping:

(i) Most of the variables which occur become n-dimensional lists. E.g.the
unknown y is a N-dimensional list as well as the critical eigenfunction CFUN
and its adjoint critical eigenfunction AFUN.

(ii) The inner product $\langle u,v \rangle$ involves a scalar product in $\mathbb{R}^N$ with subsequent
integration.

(iii) In order to keep the amount of calculation to a minimum we also assumed
again that the differential equations depend only linearly on the bifurcation
parameter α. This can, however, easily be changed by changing the iteration
range in the function REDUCTION3().

(iv) A major addition is the program called SOLVE_ODE which solves the differential equation associated with the Taylor coefficients of $w(x;A,\lambda)$ (cf. eq.(16.2)). It first identifies all trigonometric terms in a given expression using the function FIND_TRIG and the rules COSFIND, SINFIND, COSNULL and SINNULL. The function SETIFY removes multiple occurrences of the same trigonometric functions such that one can make an ansatz for a solution of the differential equation W_DE. This ansatz is then inserted into the differential equation. We assume that these trigonometric functions are the solutions of the linear eigenvalue problem for the original set of p.d.e.'s and form a complete set of orthogonal eigenfunctions on the defining space interval. Hence one can find the amplitudes of the ansatz by solving the algebraic equations which result from setting the coefficients of the trigonometric eigenfunctions to zero. The function FEEDW then guarantees that the projection $\langle y_1^*, w \rangle$ is always zero.

(v) VFUN(LIST,VALUE) is a function which creates the substitution list [LIST[1] = VALUE, LIST[2] = VALUE,...] effectively.

Liapunov-Schmidt Reduction and Amplitude Equations

In this Section we shall use the Liapunov-Schmidt reduction to treat problems which involve coupled systems of nonlinear p.d.e.'s in more than one independent variable. We begin with the description of an example.

We have dealt with the Lorenz system in Chapter 2 as well as earlier in this Chapter. This system originated from an amplitude expansion and was thought to describe the onset of two dimensional convection rolls [34]. The corresponding experimental situation is known as the 2D Rayleigh-Benard problem: A fluid is confined to a box, heated from below and cooled from above. At a certain critical temperature (manifested in the Navier-Stokes equations as a critical nondimensional number, the Rayleigh number Ra), the pure heat conductive state becomes unstable and stationary parallel convection

rolls develop. This onset of convection is conveniently described by two
partial differential equations for the stream function $\psi(x,z)$ (the
Navier–Stokes equations) and the deviation $\theta(x,z)$ of the temperature from the
purely conducting state, which are defined on a two-dimensional interval
$0 \leq x \leq L_x$, $0 \leq z \leq L_z$. Usually one makes several approximations (the
Boussinesq approximation) of which the most important ones are
incompressibility of the fluid, temperature independent material parameters and
so-called free boundary conditions which allow for simple trigonometric
solutions of the linearized problem. The governing equations for the
stationary problem become:

(44)
$$\nabla^4 \psi - d_x \theta = Pr^{-1} (d_x \psi \, d_z - d_z \psi \, d_x) \nabla^2 \psi$$

$$- \nabla^2 \theta + Ra \, d_x \psi = (d_x \psi \, d_z - d_z \psi \, d_x) \theta$$

where $\nabla^2 = d_{xx} + d_{zz}$ is the Laplace operator, and where Pr, the Prandtl number,
is a nondimensional number given by the quotient of the kinematic viscosity
coefficient and the thermal diffusivity coefficient. It turns out that for
Ra = 8,

(45)
$$\begin{bmatrix} \psi_c \\ \theta_c \end{bmatrix} = \begin{bmatrix} \sin x \sin z \\ -4 \cos x \sin z \end{bmatrix}$$

is a critical eigenfunction for the linearized problem (the left hand side of
(44)) with eigenvalue zero. With the adjoint linear operator defined by

(46)
$$L^*(Ra) = \begin{bmatrix} \nabla^4 & -Ra \, d_x \\ d_x & -\nabla^2 \end{bmatrix}$$

we find the adjoint critical eigenfunction

(47)
$$\begin{bmatrix} \psi_c^* \\ \theta_c^* \end{bmatrix} = \begin{bmatrix} \sin x \sin z \\ -\frac{1}{2} \cos x \sin z \end{bmatrix}.$$

A bifurcation analysis shows that at $Ra = Ra_c = 8$ there exists a supercritical pitchfork bifurcation to stable convecting rolls, described by the bifurcation equation $\lambda A \pi^2/8 - A^3 \pi^2/16 = 0$ with, as usual, $\lambda = Ra - Ra_c$. Supercritical means that if one varies λ and approaches the bifurcation point on the branch $A = 0$ from the stable side, then the branching solutions bifurcate in the forward direction (cf. Fig.4 in Chapter 1). Using the o.d.e. (37) it can easily be shown that supercritical solutions are stable and subcritical solutions (those branching backwards) are unstable.

In the following sample run of the Liapunov-Schmidt reduction for this problem we show how we can use slightly extended versions of the building blocks SETUP, G_POLY, W_POLY and FEEDW given in the previous section to deal with coupled partial differential equations in more than one space variable. Since it is impractical to write a general program which performs a Fourier transformation of the coupled partial differential equations which have to be solved to find $w(a;A,\lambda)$, we perform this step interactively. The solution is then given as input to the function FEEDW which puts the result into DATABASE.

```
LAPLACE(FUN) := DIFF(FUN,X,2)+DIFF(FUN,Z,2)$
Warning - you are redefining the MACSYMA function LAPLACE

SETUP();
ENTER THE NUMBER OF DIFFERENTIAL EQUATIONS
2;
ENTER THE DEPENDENT VARIABLES AS A LIST
[PSI,THETA];
ENTER NUMBER OF SPATIAL COORDINATES
2;
ENTER THE SPATIAL COORDINATES AS A LIST
[X,Z];
ENTER THE BIFURCATION PARAMETER
RA;
ENTER THE CRITICAL BIFURCATION VALUE
8;
WE DEFINE LAM =  RA - 8
ENTER THE CRITICAL EIGENFUNCTION AS A LIST
[SIN(X)*SIN(Z),-4*COS(X)*SIN(Z)];
```

ENTER THE ADJOINT CRITICAL EIGENFUNCTION AS A LIST
[SIN(X)*SIN(Z),-COS(X)*SIN(Z)/2];
ENTER THE DIFFERENTIAL EQUATION NUMBER 1
(DIFF(PSI,X)*DIFF(LAPLACE(PSI),Z)-DIFF(PSI,Z)*DIFF(LAPLACE(PSI),X))*
PRANDTL^-1-(LAPLACE(LAPLACE(PSI))-DIFF(THETA,X));
ENTER THE DIFFERENTIAL EQUATION NUMBER 2
DIFF(PSI,X)*DIFF(THETA,Z)-DIFF(PSI,Z)*DIFF(THETA,X)-RA*DIFF(PSI,X)
+LAPLACE(THETA);

ENTER THE LOWER LEFT CORNER OF THE 2 DIMENSIONAL SPACE INTERVAL
[X =...,]
  1
[X = 0,Z = 0];

ENTER THE UPPER RIGHT CORNER OF THE 2 DIMENSIONAL SPACE INTERVAL
[X =...,]
  1
[X = %PI,Z = %PI];

[SYMBOLICS 3670 TIME = 270 SEC.]

$$
\left[\frac{dTHETA}{dX} - \frac{d^4 PSI}{dZ^4} + \frac{\frac{dPSI}{dX}\left(\frac{d^3 PSI}{dZ^3} + \frac{d^3 PSI}{dX^2 dZ}\right) - \left(\frac{d^3 PSI}{dX^3} + \frac{d^3 PSI}{dX dZ^2}\right)\frac{dPSI}{dZ}}{PRANDTL} \right.
$$

$$
- \frac{d^4 PSI}{dX^4} - 2\frac{d^4 PSI}{dX^2 dZ^2},
$$

$$
\left. \frac{d^2 THETA}{dZ^2} + \frac{dPSI}{dX}\frac{dTHETA}{dZ} + \frac{d^2 THETA}{dX^2} - \frac{dPSI}{dZ}\frac{dTHETA}{dX} - (LAM + 8)\frac{dPSI}{dX}\right]
$$

G_POLY(1,0);
[SYMBOLICS 3670 TIME = 13 SEC.]

                                            0

*This is a check on whether we solved our linear problem correctly. If we are
dealing with a bifurcation problem, this coefficient is always zero.*

W_POLY(0,1);

NOW SOLVE THE EQUATIONS

$$\left[\frac{dZW2}{dX} - \frac{d^4ZW1}{dZ^4} - \frac{d^4ZW1}{dX^4} - 2\frac{d^4ZW1}{dX^2 dZ^2}, \; \frac{d^2ZW2}{dZ^2} + \frac{d^2ZW2}{dX^2} - 8\frac{dZW1}{dX}\right]$$

=0! THEY ARE GIVEN IN W_DE

[SYMBOLICS 3670 TIME = 5 SEC.]

CALL FEEDW( ) TO PROCEED

This is the linearized equation, and so the only nontrivial solution is given

by CFUN. Since <CFUN,W> = 0, we find that ZW1 = ZW2 = 0.

FEEDW( );
ENTER ZW1
0;

ENTER ZW2
0;

[SYMBOLICS 3670 TIME = 3 SEC.]

$$\left[\frac{dW1}{dLAM} = 0, \; \frac{dW2}{dLAM} = 0\right]$$

WPOLY(2,0);

NOW SOLVE THE EQUATIONS

$$\left[\frac{dZW2}{dX} - \frac{d^4ZW1}{dZ^4} - \frac{d^4ZW1}{dX^4} - 2\frac{d^4ZW1}{dX^2 dZ^2}, \right.$$

$$\left. \frac{d^2ZW2}{dZ^2} + \frac{d^2ZW2}{dX^2} - 8\frac{dZW1}{dX} + (- 4 SIN^2 (X) - 4 COS^2 (X)) SIN(2 Z)\right]$$

=0! THEY ARE GIVEN IN W_DE

[SYMBOLICS 3670 TIME = 15 SEC.]

CALL FEEDW( ) TO PROCEED

TRIGREDUCE(W_DE,X);
[SYMBOLICS 3670 TIME = 1 SEC.]

$$\left[\frac{dZW2}{dX} - \frac{d^4ZW1}{dZ^4} - \frac{d^4ZW1}{dX^4} - 2\frac{d^4ZW1}{dX^2 dZ^2}, \; \frac{d^2ZW2}{dZ^2} + \frac{d^2ZW2}{dX^2} - 8\frac{dZW1}{dX} - 4 SIN(2 Z)\right]$$

```
EV(%,ZW1 = AMP1*SIN(2*Z),ZW2 = AMP2*SIN(2*Z),DIFF);
[SYMBOLICS 3670 TIME = 1 SEC.]
```

$$[- 16 \text{ AMP1 SIN}(2\ Z),\ -\ 4\ \text{AMP2 SIN}(2\ Z) - 4\ \text{SIN}(2\ Z)]$$

```
FEEDW();
ENTER ZW1
0;

ENTER ZW2
-SIN(2*Z);

[SYMBOLICS 3670 TIME = 13 SEC.]
```

$$\left[\frac{d^2\ W1}{dAMP^2} = 0,\ \frac{d^2\ W2}{dAMP^2} = -\ \text{SIN}(2\ Z)\right]$$

```
BIFEQ:G_POLY(1,1)*AMP*LAM+G_POLY(3,0)*AMP^3/3!;

[SYMBOLICS 3670 TIME = 1 SEC.]
```

$$\frac{\%PI\ AMP^2\ LAM}{8} - \frac{\%PI\ AMP^2\ ^3}{16}$$

Since the program largely duplicates the corresponding functions given in the the previous Section, we do not supply a complete listing of the functions used here, but rather we only discuss some important new features.

The major difference between the programs used here and those given in the previous Section is that we are dealing with more than one independent variable, say SPACE variables. These have to be entered into a list, say XVAR, and the dependency of the independent variables Y and the variables W and ZW on XVAR has to be declared. Also the projections P and Q now involve integration over several space intervals. This is most effectively performed via the following auxiliary function. (Do not use multiple definite integration in MACSYMA. It is very slow and keeps asking you annoying questions about the sign of totally uninteresting quantities):

```
INT(EXP):=(INTINT:EXP,

            FOR I THRU SPACE DO

                INTINT:INTEGRATE(TRIGREDUCE(INTINT,XVAR[I]),XVAR[I]),

            RATSIMP(EV(INTINT,UBOUND) - EV(INTINT,LBOUND)))$
```

where LBOUND is the lower left corner of the SPACE-dimensional simplex given as
a list $[X_1= \quad ,X_2= \quad ....]$ and UBOUND is the corresponding upper right corner.

What has all this to do with the Lorenz equations? We observe from the
last demonstration run that for the steady state bifurcation of the Navier
Stokes equations there are only three relevant modes for a bifurcation equation
up to third order: the two modes of the critical eigenfunction sin x sin z and
cos x sin z and the perturbation mode in the temperature equation sin 2z. If
one makes the ansatz

$$[\psi,\theta] = [\ A \sqrt{8} \sin x \sin z\ , \ -4 \sqrt{8} B \cos x \sin z\ - 8 C \sin 2z\ ]$$

we can recover the Lorenz model, as we show next:

EQS;

```
             2    2       2
   dTHETA   d    d PSI   d PSI
  [------- - --- (----- + -----)
     dX      2     2       2
            dZ    dZ      dX
```

```
            2       2             2       2
  dPSI  d  d PSI   d PSI    dPSI  d  d PSI   d PSI
  ---- (-- (----- + -----)) - ---- (-- (----- + -----))
  dX    dZ    2       2       dZ    dX    2       2
            dZ      dX                   dZ      dX
+ -----------------------------------------------------
                        PRANDTL
```

```
                              2       2
                             d    d PSI   d PSI
                             -- (----- + -----)
       2       2             dT    2       2
      d    d PSI   d PSI         dZ      dX
  - --- (----- + -----)  +  --------------------- ,
      2     2       2              PRANDTL
    dX    dZ      dX
```

```
   2                              2
  d THETA   dPSI  dTHETA     d THETA   dPSI  dTHETA    THETA          dPSI
  ------- + ---- ------- +  ------- - ---- ------- - ------ - RA  ----]
     2       dX    dZ          2       dZ    dX       dT           dX
   dZ                        dX
```

EV(EQS,PSI = A*2*2^(1/2)*CFUN[1],
THETA = B*2*2^(1/2)*CFUN[2]-C*8*SIN(2*Z),DIFF)$

TRIGREDUCE(%,X)$

TRIGREDUCE(%,Z);

```
                 dA
        4 SQRT(2) -- SIN(X) SIN(Z)
                 dT
  [- ------------------------------- + 8 SQRT(2) B SIN(X) SIN(Z)
              PRANDTL

   - 8 SQRT(2) A SIN(X) SIN(Z),

                              SIN(3 Z)   SIN(Z)
      2 SQRT(2) A (- 16 C COS(X) (-------- - ------)
                                    2         2

   - 2 SQRT(2) B COS(2 X) SIN(2 Z) - 2 SQRT(2) B SIN(2 Z))

              1   COS(2 X)            dC
   - 16 A B (- - --------) SIN(2 Z) + 8 -- SIN(2 Z) + 32 C SIN(2 Z)
              2      2                 dT

                                              dB
   - 2 SQRT(2) A RA COS(X) SIN(Z) + 8 SQRT(2) -- COS(X) SIN(Z)
                                              dT

   + 16 SQRT(2) B COS(X) SIN(Z)]
```

SOLVE([RATCOEF(%,COS(X)*SIN(Z))[2],RATCOEF(%,SIN(X)*SIN(Z))[1],
RATCOEF(%,SIN(2*Z))[2]],[DIFF(A,T),DIFF(B,T),DIFF(C,T)]);

```
    dA                               dB   A RA - 8 A C - 8 B
  [[-- = 2 B PRANDTL - 2 A PRANDTL,  -- = ------------------,
    dT                               dT           4

    dC
    -- = 2 A B - 4 C]]
    dT
```

Note that with rescaling time d/dt → 2( )' and setting β = 2, RA = 8ρ,
the amplitude equations for A,B,C are identical to the Lorenz equations (35).
Even the bifurcation equation for the Liapunov-Schmidt reduction from the
Navier-Stokes equations and the reduction starting with the Lorenz model are
identical for β = 2.

Whereas most of the amplitude expansions which are popular especially in hydrodynamics seem to be constructed on physical arguments, we can offer here a rigorous procedure to determine the necessary modes for an ansatz (cf.[15]): If we want to be exact up to a certain order, we have to take into account all the modes which occur in the Liapunov-Schmidt reduction up to this order. Note that this does not require that we actually perform the reduction. It is much simpler to find the relevant modes than to perform the reduction, especially since it does not involve solving the differential equation (16.2) but only orthogonal expansions as in the foregoing sample run (cf. also the following Section). Hence sometimes it pays to do a "reduced" Liapunov-Schmidt reduction, i.e., determining only the relevant modes and then making an amplitude expansion with respect to these modes. Then one can use such powerful techniques as normal form transformations or center manifold reductions on the set of modal equations. These are o.d.e.'s and hence involve only polynomial manipulations for which MACSYMA is much better equipped than for dealing with p.d.e.'s. A last remark: Obviously these modal expansions cannot say anything more about the true solutions of the p.d.e. than the corresponding Liapunov-Schmidt reduction. In particular all solutions are valid only locally and are isomorphic to the true solutions up to the degree for which the modes of the amplitude expansion correspond to the modes in the Liapunov-Schmidt reduction. Hence the chaotic solutions for which the Lorenz system is famous are not directly related to the solutions of the Navier-Stokes equations.

Determination of Higher Order Coefficients

If we try to find the bifurcation equation for some nontrivial partial differential equations up to higher order we may run into trouble. For instance the intermediate expressions which one encounters using the function G_POLY and W_POLY on the Benard problem to determine a fifth order coefficient are quite large. This situation becomes even worse when we think of calculating bifurcation equations up to higher order for degenerate problems, i.e. bifurcations with more than one critical eigenfunction becoming simultaneously unstable. In principle the procedure is straightforward: For a problem with n critical eigenfunctions we have an n-dimensional kernel N which is spanned by $A_1 u_{1c}$, $A_2 u_{2c}$, ...$A_n u_{nc}$, the n critical eigenfunctions and their amplitudes. The projection P onto the kernel leads now to n coupled algebraic bifurcation equations $g_1(A_1, ..., A_n, \lambda)$, ..., $g_n(A_1, ..., A_n, \lambda)$ and the function w which lies in the complement of N is also a function of n amplitudes $w(x; A_1, ..., A_n, \lambda)$. Hence it is conceptually not difficult to extend the function G_POLY and W_POLY to the n-fold degenerate steady state bifurcation problem. However, due to intermediate expression swell, this program will be of very limited use for coefficients higher than third order. Also we encounter one of the basic issues of Computer Algebra: Even if it is possible to calculate nontrivial fifth order coefficients, these are usually large and complicated algebraic expressions, depending on the parameters of the problem. In general these expressions are too difficult to interpret and hence are meaningless. However in some interesting cases we have been able to reduce the number of coefficients involved by fixing them at critical values such that lower order terms in the bifurcation equations vanish. Then the next higher order terms are the leading order terms and therefore their signs determine the whole local bifurcation structure. If we are lucky then the relevant coefficients depend only on one or two parameters and can be plotted as functions of these parameters.

The following interactive procedure describes a highly efficient way of calculating these higher order coefficients and plotting their graphs. Its key requirement is that products of the critical eigenfunctions must have a finite decomposition into a complete set of suitable orthogonal eigenfunctions. For trigonometric eigenfunctions this means that we need such nonlinear terms that the Fourier decomposition of the occurring products of the critical eigenfunctions are finite. The aim of the procedure is to create a FORTRAN program out of MACSYMA which eventually feeds into a plotting routine.

Rather than presenting the procedure in full generality, we treat the Benard problem of (44) as an example (cf.[1],[21]). We write the Boussinesq equations as

$$(48) \qquad\qquad L \begin{bmatrix} \psi \\ \theta \end{bmatrix} = NL\ (\psi,\theta)$$

where ψ,θ are the stream function and the temperature variation respectively, L is the linear operator and NL (ψ,θ) a vector-valued nonlinear differential operator. For a 2-dimensional infinitely extended problem and "free" horizontal boundary conditions at $z = -1/2$ and $1/2$, we find two simultaneously unstable critical modes at a critical periodicity length defined by α:

$$\psi_c = [A\ \sin(\alpha x+\varphi) + B\ \sin(2\alpha x+\chi)]\ \sin\ \pi(z+\tfrac{1}{2})$$

$$(49) \qquad\qquad \psi_c \equiv A\ \psi^{(1)} + B\ \psi^{(2)}$$

$$\theta_c = -\ [A\ (\pi^2+\alpha^2)^2\ \alpha^{-1}\ \cos(\alpha x+\varphi) + B\ (\pi^2+4\alpha^2)^2\ (2\alpha)^{-1}\ \cos(2\alpha x+\chi)]\ \sin\ \pi(z+\tfrac{1}{2})$$

$$(50) \qquad\qquad \theta_c \equiv A\ \theta^{(1)} + B\ \theta^{(2)}.$$

For this degeneracy the critical Rayleigh number Ra_c and the critical wavelength α are given by $Ra_c = 769.234$, $\alpha = 1.5501534$. In this way we model a very large Benard problem (length of container $\gg 1/\alpha$) as an infinitely extended problem with periodic boundary conditions. The critical wavevector α then corresponds to a periodicity length L such that the modes α and 2α become simultaneously unstable. Such an L can always be found. Since we impose only periodic boundary conditions we allow for arbitrary variations in the phases φ and χ.

Symmetry arguments [1] reveal that the generic system for this problem is of the form

$$A' = (\lambda + a_{11} A^2 + a_{12} B^2) A + c_1 A^3 B^2 \cos 2\phi$$

(51)
$$B' = (\lambda + a_{21} A^2 + a_{22} B^2) B + c_2 A^4 B \cos 2\phi$$

$$\phi' = - A^2 (2c_1 B^2 + c_2 A^2) \sin 2\phi$$

where $\phi = 2\varphi - \chi$. The first occurrence of the phase ϕ in the bifurcation equations is in fifth order terms. These terms are crucial for the solution structure of the bifurcation equations (51). To determine them we make the following ansatz:

$$\psi = \psi_c + w\psi_1 + w\psi_2 + \ldots$$

(52)
$$\theta = \theta_c + w\theta_1 + w\theta_2 + \ldots$$

where $w\psi_i, w\theta_i = O(A^{i+1}, B^{i+1})$. Clearly in order to determine a fifth order term in the bifurcation equation for a p.d.e. with quadratic nonlinearity we have to know the function w, i.e. here $w\psi$ and $w\theta$, to order four in A and B. The key observation to reducing the necessary calculations is that, in order to determine a specific fifth order coefficient, we do not need w up to fifth order exactly. We only need to know certain Fourier coefficients of it. We proceed stepwise:

Step 1: Insert (52) into (48) and expand to order 2 in A and B. This gives:

$$(53) \qquad L \begin{bmatrix} w\psi_1 \\ w\theta_1 \end{bmatrix} = NL\ (\psi_c, \theta_c)$$

since $L \begin{bmatrix} \psi_c \\ \theta_c \end{bmatrix} = 0$ by definition. Now we use the command TRIGREDUCE in MACSYMA on the right hand side of equation (53) to give $NL\ (\psi_c, \theta_c)$ decomposed into its Fourier modes, say

$$f_1 \text{trig } w\psi_1^1 \ \ldots \ f_n \text{trig } w\psi_1^n \text{ and } g_1 \text{trig } w\theta_1^1 \ \ldots \ g_m \text{trig } w\theta_1^m$$

Here trig $w\psi_k^j$ and trig $w\theta_k^1$ with $1 \leq j \leq n$, $1 \leq 1 \leq m$ stands for one of the n and m possible Fourier modes in $w\psi_k$ and $w\theta_k$ (cf. eq.(52)), respectively. E.g. we find for the Boussinesq eq.(48)

$$w\psi_1^1 = \sin(\alpha x + \chi - \varphi)\ \sin 2\pi(z + \tfrac{1}{2}) \quad \text{and} \quad w\theta_2^1 = \cos(\chi - 2\varphi)\ \sin \pi(z + \tfrac{1}{2}).$$

The coefficients f_i and g_i depend on A or B or both. Now we set

$$w\psi_1 = c_1^1 \text{ trig } w\psi_1^1 + \ldots \quad c_1^n \text{ trig } w\psi_1^n$$

$$(54)$$

$$w\theta_1 = d_1^1 \text{ trig } w\theta_1^1 + \ldots \quad d_1^m \text{ trig } w\theta_1^m$$

where the coefficients c_1^i, d_1^j are as yet unknown functions of O(2) in A and B.

which have the same argument as the corresponding coefficients f_i, g_j. We do not solve (53) in this step.

     Step 2: Insert (52) with $(w\psi_1, w\theta_1)$ given by (54) into (48) and expand to order 3 in A and B. This gives:

$$(55) \qquad L \begin{bmatrix} w\psi_2 \\ w\theta_2 \end{bmatrix} = \widetilde{NL}\ (w\psi_1, w\psi_c) + NL\ (w\psi_1, w\theta_c) + NL\ (w\psi_c, w\theta_1)$$

where $\widetilde{NL}$ is that part of the Boussinesq equations which is quadratic in ψ (cf.(44)). We convert the right hand side again to Fourier modes whose coefficients now depend on $A, B, c_1^1, \ldots, c_1^n, d_1^1, \ldots, d_1^m$ and set

$$
\begin{aligned}
w\psi_2 &= c_2^1 \text{ trig } w\psi_2^1 + \ldots + c_2^v \text{ trig } w\psi_2^v \\
(56) \\
w\theta_2 &= d_2^1 \text{ trig } w\theta_2^1 + \ldots + d_2^s \text{ trig } w\theta_2^s
\end{aligned}
$$

where v and s are the number of Fourier modes in $w\psi_2$ and $w\theta_2$ respectively. Again, the c_2^i, d_2^j have the same functional dependence on the above set of parameters as the corresponding coefficients of the r.h.s. of (55). We proceed in this way to order 5 and project the nonlinear part of O(5) onto the eigenspaces given by $\begin{bmatrix} w\psi(1) \\ w\theta(1) \end{bmatrix}$ and $\begin{bmatrix} w\psi(2) \\ w\theta(2) \end{bmatrix}$. In each of the resulting algebraic equations there is only one term which depends on the phases φ and χ, namely $\cos 2\varphi - \chi$, which are the coefficients c_1 and c_2 of (51), respectively. They depend on the previous coefficients c_3^i, d_3^j, c_2^k, ... which in turn determine the crucial modes trig $w\psi_3^i$, trig $w\theta_3^j$, trig $w\psi_2^k$, ... relevant for the fifth order terms c_1 and c_2. In order to evaluate c_1 and c_2, we trace back the above calculation and determine in each step the inverse L^{-1} for these critical modes. In this way we find the coefficients c_i^k as functions of the coefficients c_{i-1}^j and d_{i-1}^1. However we solve each step separately such that

we never accumulate large algebraic expressions. Instead we write the results
out from MACSYMA into FORTRAN and finally plot a graph $c_1 = c_1(Pr)$,
$c_2 = c_2(Pr)$. Our example shows the efficiency of this method: There are of the
order of 100 modes in the function w which contribute to fifth order. However
for the coefficients c_1 and c_2 the only relevant modes are:

$$w\psi_1 \simeq \sin(\alpha x + \chi - \varphi) \sin 2\pi(z + \tfrac{1}{2})$$

$$w\psi_2 \simeq [\sin(\chi - 2\varphi) + \sin(3\alpha x + 2\chi - \varphi)] \sin \pi(z + \tfrac{1}{2})$$

$$+ [\sin(\chi - 2\varphi) + \sin(3\alpha x + 2\chi - \varphi)] \sin 3\pi(z + \tfrac{1}{2})$$

$$w\psi_3 \simeq [\sin(\alpha x - \chi + 3\varphi) + \sin(2\alpha x + 2\chi - 2\varphi)] \sin 2\pi(z + \tfrac{1}{2})$$

and corresponding cosine modes for $w\theta_i$, $i = 1,2,3$.

Exercises

1. Change the program REDUCTION1 such that it runs correctly and most efficiently for steady state bifurcations from o.d.e.'s. Reproduce the results of the sample run of REDUCTION2 on the Lorenz equations (35).

Hint: You are dealing with purely algebraic equations. Here the differential equation (16.2) reduces to a system of nonlinear algebraic equations. The resulting equations for the Taylor coefficients of $w(x;A,\lambda)$ are systems of linear equations which can easily be solved automatically.

2. A variant of the problem of eqs. (39),(40) is the following integrodifferential equation [6] which describes the motion of an elastic rod with hinged ends.

$$y_1{}_t = y_2$$

(P1)

$$y_2{}_t = -y_1{}_{xxxx} + \alpha\, y_1{}_{xx} - y_2 + \frac{2}{\pi^4} y_1{}_{xx} \int_0^1 (y_1{}_\xi(\xi))^2 d\xi$$

with $y_1 = y_2 = 0$ at $x = 0,1$. Use the program REDUCTION3 to show that the bifurcation equation up to fifth order for (P1) is given by

$$-\pi^2\,\lambda A - 4\pi^2\,\lambda A^3 - A^3 - 3A^5 = 0$$

3. The semilinear wave equation (39) can be treated in an alternate fashion:
Since we are only interested in stationary bifurcations, we can perform the
Liapunov–Schmidt reduction on the one-dimensional differential equation

(P2) $- v_{xx} - \alpha v + v^3 + \rho v^5 = 0$

with $v = 0$ at $x = 0, \pi$. Use the program REDUCTION1 to determine the bifurcation
equation up to fifth order for (P2). Compare your result with the result
obtained in the text by using REDUCTION3.

The differences can be understood in the following way: $\left. \dfrac{d^3 w}{dA^3} \right|_{\substack{A=0 \\ \lambda=0}}$ for the

two dimensional version is proportional to $\begin{bmatrix} c \sin x + d \sin 3x \\ - c \sin x \end{bmatrix}$, whereas for

the one dimensional version $\left. \dfrac{d^3 w}{dA^3} \right|_{\substack{A=0 \\ \lambda=0}}$ is proportional only to sin 3x. Hence the

fifth order coefficient in the bifurcation equation in the former case contains

an additional term of the form $\int_0^\pi \sin x \, \rho \, \sin^2 x \, \dfrac{d^3 w}{dA^3} \, dx \simeq \int_0^\pi \rho \, \sin^4 x \, dx \neq 0$.

This is a good example of the nonuniqueness of the Liapunov–Schmidt
reduction. Note however that the local information for λ, A near zero is the
same in both cases. This is manifested by the solutions of the bifurcation
equation which are $\lambda = 3/4 \, A^2$ + terms of higher order, where the two reductions
differ only by these higher order terms. For a more extensive discussion of
the equivalence of different Liapunov–Schmidt reductions and the question where
to truncate a bifurcation equation, see [13].

4. Calculate the bifurcation equation for the one-dimensional problem

(P3) $y'' + \alpha y + y^2 = 0$

with Dirichlet boundary conditions $y(0) = y(1) = 0$.

For $\alpha = \pi^2$ the trivial solution is unstable against a perturbation of

the form $\sin \pi x$. Note that the differential equation for $\dfrac{d^2 w}{dA^2}\Big|_{\substack{A=0 \\ \lambda=0}}$ ($= zw$)

becomes

(P4) $zw'' + \pi^2 zw = - 2 \sin^2\pi x = - (1 - \cos 2\pi x)$.

If we try to solve this equation via a Fourier transformation with those

Fourier modes that satisfy the boundary conditions (i.e. a pure sine-series),

we find that the solution zw is given by an infinite series. The approach in

REDUCTION1 avoids this by solving the o.d.e.(P4) directly, introducing a

"resonant" term of the form $x \sin \pi x$. Check that the solution for $\dfrac{d^2 w}{dA^2}\Big|_{\substack{A=0 \\ \lambda=0}}$ does

indeed satisfy the boundary conditions and is orthogonal to $\sin \pi x$. Can you

think of a way of extending this approach to n-dimensional systems, n>1?

5. The Brusselator is a well–known example of a reaction diffusion equation [33]. Its steady state bifurcations from a homogeneous distribution of two chemical substances y_1 and y_2 is described by

$$(- d_1 d_{xx} - b + 1) y_1 - a^2 y_2 - NL(y_1, y_2) = 0$$

(P5)

$$(- d_2 d_{xx} + a^2) y_2 + b y_1 + NL(y_1, y_2) = 0$$

where

$$NL(y_1, y_2) = - (b/a \, y_1^2 + 2a \, y_1 \, y_2 + y_1^2 \, y_2)$$

and where a and b are externally controllable chemical substances and $d_1 d_{xx}$ and $d_2 d_{xx}$ describe the diffusion. Usually one treats b as the bifurcation parameter. Let us define the system on the interval $[0, \pi]$ with Dirichlet boundary conditions.

Show that for $a^2 = d_1 d_2$, (P5) has a zero eigenvalue at $b = (1 + d_1)^2$ with critical eigenfunction $\sin x \begin{bmatrix} d_2 \\ - 1 - d_1 \end{bmatrix}$ and adjoint critical eigenfunction $\sin x \begin{bmatrix} 1 + 1/d_1 \\ 1 \end{bmatrix}$. Find the bifurcation equation up to second order. Determine why the program SOLVE_ODE encounters inconsistent equations when attempting to calculate W_POLY(2,0) (cf. exercise 4 and [4]). Show that for $d_1 = 1$ the first nonvanishing term in the bifurcation equation is of third order and calculate this term.

APPENDIX

INTRODUCTION TO MACSYMA

Introduction

The purpose of this Appendix is to provide an introduction to the
computer algebra system MACSYMA. We have included this introduction so that
the book would be understandable to the reader with no familiarity with
MACSYMA. However, it is by necessity so brief that the reader is referred to
other sources for further information ([26],[35]).

If, as we intend, the reader plans to use our MACSYMA programs to
perform perturbation calculations, it is unquestionably desirable that some
first-hand experience with MACSYMA be available before trying our programs.
For a reader with no MACSYMA experience, we suggest reading this introduction
in front of a computer terminal running MACSYMA. Then the reader may wish to
duplicate our run, punctuated with explorations of the reader's design. At the
end of the Appendix we offer some elementary exercises for the beginning
MACSYMA user.

To invoke MACSYMA on most machines, type:

MACSYMA (return)

The computer will display a greeting of the sort:

This is MACSYMA 304

[copyright message ...]

Loading fix file <MACSYM>TOPS20.FIX.304

(C1)

The (C1) is a "label". Each input or output line is labelled and can be
referred to by its label for the rest of the session. C labels denote your
Commands and D labels denote Displays of the machine's response. Never use
variable names like C1 or D5, as these will be confused with the lines so
labeled.

Special Keys and Symbols

1. To end a MACSYMA session, type QUIT(); . On some machines ^C or ^Y or ^Z
will also work. (Here ^ stands for the control key, so that ^C means
simultaneously press both the key marked control and the C key).

2. To abort a computation without leaving MACSYMA, type ^G or ^C or press the
ABORT key. This step will differ from machine to machine. It is important for
you to know how to do this on your machine, in case, for example, you begin a
computation which is taking too long.

3. In order to tell MACSYMA that you have finished your command, use the
semicolon (;). On some machines you must follow the ; with a return. Note
that the return key alone does not signal that you are done with your input.

4. An alternative input terminator to the semicolon (;) is the dollar sign ($),
which, however, supresses the display of MACSYMA's computation. This is useful
if you are computing some long intermediate result, and you don't want to waste
time having it displayed on the screen.

5. If you want to completely delete the current input line (and start this line fresh from the beginning), type a double question mark (??).

6. If you wish to repeat a command which you have already given, say on line (C5), you may do so without typing it over again by preceding its label with a double quote (''), i.e., ''C5. (Note that simply inputing C5 will not do the job - try it.)

7. If you want to refer to the immediately preceding result computed by MACSYMA, you can either use its D label, or you can use the special symbol percent (%).

8. The standard quantities e (natural log base), i (square root of -1) and pi (3.14159...) are respectively referred to as %E, %I and %PI. Note that the use of % here as a prefix is completely unrelated to the use of % to refer to the preceding result computed.

9. In order to assign a value to a variable, MACSYMA uses the colon (:), not the equal sign. The equal sign is used for representing equations.

Arithmetic

The common arithmetic operations are:

+ addition

- subtraction

* scalar multiplication

/ division

^ or ** exponentiation

. matrix multiplication

! factorial

SQRT(X) square root of X

MACSYMA's output is characterized by exact (rational) arithmetic. E.g.,

(C1) 1/100+1/101;

$$(D1) \qquad \frac{201}{10100}$$

If irrational numbers are involved in a computation, they are kept in symbolic form:

(C2) (1+SQRT(2))^5;

$$(D2) \qquad (SQRT(2) + 1)^5$$

(C3) EXPAND(%);
(D3)   29 SQRT(2) + 41

However, it is often useful to express a result in decimal notation. This may be accomplished by following the expression you want expanded by ",NUMER":

(C4) %,NUMER;
(D4)   82.012194

Note the use here of % to refer to the previous result. In this version of MACSYMA, NUMER only gives 8 significant figures, of which the last is often unreliable. However, MACSYMA can offer <u>arbitrarily high precision</u> by using the BFLOAT function:

(C5) BFLOAT(D3);
(D5)   8.2012193308819768B1

The number of significant figures displayed is controlled by the MACSYMA variable FPPREC, which has the default value of 16:

(C6) FPPREC;
(D6)   16

Here we reset FPPREC to yield 100 digits:

(C7) FPPREC:100;
(D7)   100
(C8) ''C5;
(D8) 8.201219330881975641524897300208124427852048438593149412212371240173124#
41875401104126661238495501605618B1

Note the use of the double quote ('') in (C8) to repeat command (C5). MACSYMA

can handle very large numbers without approximation:

(C9) 100!;
(D9) 93326215443944152681699238856266700490715968264381621468592963895217597#
99932299156089414639761565182862536979208272237582511852109168640000000000000#
0000000000000

Algebra

MACSYMA's importance as a computer tool to facilitate perturbation calculations

becomes evident when we see how easily it does algebra for us. Here's an

example in which a polynomial is expanded:

(C1) (X+3*Y+X^2*Y)^3;
$$\text{(D1)}\qquad\qquad (X^2\ Y + 3\ Y + X)^3$$
(C2) EXPAND(%);

$$\text{(D2)}\quad X^6\ Y^3 + 9\ X^4\ Y^3 + 27\ X^2\ Y^3 + 27\ Y^3 + 3\ X^5\ Y^2$$

$$+\ 18\ X^3\ Y^2 + 27\ X\ Y^2 + 3\ X^4\ Y + 9\ X^2\ Y + X^3$$

Now suppose we wanted to substitute 5/Z for X in the above expression:

(C3) D2,X=5/Z;

$$\text{(D3)}\quad \frac{135\ Y^2}{Z} + \frac{675\ Y^3}{Z^2} + \frac{225\ Y}{Z^2} + \frac{2250\ Y^2}{Z^3} + \frac{125}{Z^3} + \frac{5625\ Y^3}{Z^4}$$

$$+\ \frac{1875\ Y^2}{Z^4} + \frac{9375\ Y^3}{Z^5} + \frac{15625\ Y^3}{Z^6} + 27\ Y^3$$

The MACSYMA function RATSIMP will place this over a common denominator:

(C4) RATSIMP(%);

```
       3 6          2 5           3           4
(D4) (27 Y  Z  + 135 Y  Z  + (675 Y  + 225 Y) Z

       2            3           3          2      2
+ (2250 Y  + 125) Z  + (5625 Y  + 1875 Y) Z  + 9375 Y  Z

       3 6
+ 15625 Y )/Z
```

Expressions may also be FACTORed:

(C5) FACTOR(%);

$$(D5) \qquad \frac{(3 Y Z^2 + 5 Z + 25 Y)^3}{Z^6}$$

MACSYMA can obtain exact solutions to systems of nonlinear algebraic equations.
In this example we SOLVE three equations in the three unknowns A,B,C:

(C6) A+B*C=1;
(D6)   B C + A = 1
(C7) B-A*C=0;
(D7)   B - A C = 0
(C8) A+B=5;
(D8)   B + A = 5

(C9) SOLVE([D6,D7,D8],[A,B,C]);

$$(D9) \quad [[A = \frac{25 \text{ SQRT}(79) \text{ \%I} + 25}{6 \text{ SQRT}(79) \text{ \%I} - 34}, \ B = \frac{5 \text{ SQRT}(79) \text{ \%I} + 5}{\text{SQRT}(79) \text{ \%I} + 11},$$

$$C = \frac{\text{SQRT}(79) \text{ \%I} + 1}{10}], \ [A = \frac{25 \text{ SQRT}(79) \text{ \%I} - 25}{6 \text{ SQRT}(79) \text{ \%I} + 34},$$

$$B = \frac{5 \text{ SQRT}(79) \text{ \%I} - 5}{\text{SQRT}(79) \text{ \%I} - 11}, \ C = - \frac{\text{SQRT}(79) \text{ \%I} - 1}{10}]]$$

Note that the display consists of a "list", i.e., some expression contained
between two brackets [•••], which itself contains two lists. Each of the
latter contain a distinct solution to the simultaneous equations.

Trig identities are easy to manipulate in MACSYMA. The function TRIGEXPAND uses the sum-of-angles formulas to make the argument inside each trig function as simple as possible:

(C10) SIN(U+V)*COS(U)^3;

$$\text{(D10)} \quad COS^3(U) \; SIN(V + U)$$

(C11) TRIGEXPAND(%);

$$\text{(D11)} \quad COS^3(U) \; (COS(U) \; SIN(V) + SIN(U) \; COS(V))$$

The function TRIGREDUCE, on the other hand, converts an expression into a form which is a sum of terms, each of which contains only a single SIN or COS:

(C12) TRIGREDUCE(D10);

$$\text{(D12)} \quad \frac{SIN(V + 4\,U) + SIN(V - 2\,U)}{8} + \frac{3\,SIN(V + 2\,U) + 3\,SIN(V)}{8}$$

The functions REALPART and IMAGPART will return the real and imaginary parts of a complex expression:

(C13) W:3+K*%I;

$$\text{(D13)} \quad \%I\,K + 3$$

(C14) W^2*%E^W;

$$\text{(D14)} \quad (\%I\,K + 3)^2 \; \%E^{\%I\,K + 3}$$

(C15) REALPART(%);

$$\text{(D15)} \quad \%E^3 \; (9 - K^2) \; COS(K) - 6\,\%E^3 \; K\,SIN(K)$$

Calculus

MACSYMA can compute derivatives and integrals, expand in Taylor series, take
limits, and obtain exact solutions to ordinary differential equations. We
begin by defining the symbol F to be the following function of X:

(C1) F:X^3*%E^(K*X)*SIN(W*X);

$$\text{(D1)} \qquad X^3 \; \%E^{K\,X} \; SIN(W\,X)$$

We compute the derivative of F with respect to X:

(C2) DIFF(F,X);

$$\text{(D2)} \quad K\,X^3\; \%E^{K\,X}\; SIN(W\,X) + 3\,X^2\; \%E^{K\,X}\; SIN(W\,X)$$

$$+ W\,X^3\; \%E^{K\,X}\; COS(W\,X)$$

Now we find the indefinite integral of F with respect to X:

(C3) INTEGRATE(F,X);

$$\text{(D3)} \; (((K\,W^6 + 3\,K^3\,W^4 + 3\,K^5\,W^2 + K^7)\,X^3$$

$$+ (3\,W^6 + 3\,K^2\,W^4 - 3\,K^4\,W^2 - 3\,K^6)\,X^2$$

$$+ (-18\,K\,W^4 - 12\,K^3\,W^2 + 6\,K^5)\,X - 6\,W^4 + 36\,K^2\,W^2$$

$$- 6\,K^4)\,\%E^{K\,X}\; SIN(W\,X) + ((-W^7 - 3\,K^2\,W^5 - 3\,K^4\,W^3$$

$$- K^6\,W)\,X^3 + (6\,K\,W^5 + 12\,K^3\,W^3 + 6\,K^5\,W)\,X^2$$

$$+ (6\,W^5 - 12\,K^2\,W^3 - 18\,K^4\,W)\,X - 24\,K\,W^3 + 24\,K^3\,W)\,\%E^{K\,X}$$

$$COS(W\,X))/(W^8 + 4\,K^2\,W^6 + 6\,K^4\,W^4 + 4\,K^6\,W^2 + K^8)$$

A slight change in syntax gives definite integrals:

(C4) INTEGRATE(1/X^2,X,1,INF);
(D4)   1
(C5) INTEGRATE(1/X,X,0,INF);
INTEGRAL IS DIVERGENT

Next we define the symbol G in terms of F (previously defined in C1) and the

hypebolic sine function, and find its Taylor series expansion (up to, say,

order 3 terms) about the point X=0:

(C6) G:F/SINH(K*X)^4;

$$
(D6) \qquad \frac{X^3 \, \%E^{K X} \, SIN(W X)}{SINH^4(K X)}
$$

(C7) TAYLOR(G,X,0,3);

$$
(D7)/T/ \quad \frac{W}{K^4} + \frac{W X}{K^3} - \frac{(W K^2 + W^3) X^2}{6 K^4} - \frac{(3 W^2 K^3 + W^3) X^3}{6 K^3} + \ldots
$$

The limit of G as X goes to 0 is computed as follows:

(C8) LIMIT(G,X,0);

$$
(D8) \qquad \frac{W}{K^4}
$$

MACSYMA also permits derivatives to be represented in unevaluated form (note

the quote):

(C9) 'DIFF(Y,X);

$$
(D9) \qquad \frac{dY}{dX}
$$

The quote operator in (C9) means "do not evaluate". Without it, MACSYMA would

have obtained 0:

(C10) DIFF(Y,X);
(D10)   0

Using the quote operator we can write differential equations:

(C11) 'DIFF(Y,X,2)+'DIFF(Y,X)+Y;

$$
\text{(D11)} \qquad \frac{d^2 Y}{dX^2} + \frac{dY}{dX} + Y
$$

MACSYMA's ODE2 function can solve some first and second order ODE's:

(C12) ODE2(D11,Y,X);

$$
\text{(D12)} \quad Y = \%E^{-X/2} \left(\%K1 \, SIN\left(\frac{SQRT(3)\,X}{2}\right) + \%K2 \, COS\left(\frac{SQRT(3)\,X}{2}\right) \right)
$$

Matrix Calculations

MACSYMA can compute the determinant, inverse and eigenvalues and eigenvectors

of matrices which have symbolic elements (i.e., elements which involve

algebraic variables.) We begin by entering a matrix M element by element:


```
(C1) M:ENTERMATRIX(3,3);
Is the matrix  1. Diagonal  2. Symmetric  3. Antisymmetric  4. General
Answer 1, 2, 3 or 4
4;
Row 1 Column 1:  0;
Row 1 Column 2:  1;
Row 1 Column 3:  A;
Row 2 Column 1:  1;
Row 2 Column 2:  0;
Row 2 Column 3:  1;
Row 3 Column 1:  1;
Row 3 Column 2:  1;
Row 3 Column 3:  0;
Matrix entered.
```

$$
\text{(D1)} \qquad \begin{bmatrix} 0 & 1 & A \\ 1 & 0 & 1 \\ 1 & 1 & 0 \end{bmatrix}
$$

Next we find its transpose, determinant and inverse:

(C2) TRANSPOSE(M);

$$
\begin{bmatrix}
0 & 1 & 1 \\
1 & 0 & 1 \\
A & 1 & 0
\end{bmatrix}
$$

(D2)

(C3) DETERMINANT(M);
(D3) $A + 1$

(C4) INVERT(M),DETOUT;

$$
\dfrac{\begin{bmatrix}
-1 & A & 1 \\
1 & -A & A \\
1 & 1 & -1
\end{bmatrix}}{A + 1}
$$

(D4)

In (C4), the modifier DETOUT keeps the determinant outside the inverse. As a check, we multiply M by its inverse (note the use of the period to represent matrix multiplication):

(C5) RATSIMP(M.D4);

$$
\begin{bmatrix}
1 & 0 & 0 \\
0 & 1 & 0 \\
0 & 0 & 1
\end{bmatrix}
$$

(D5)

In order to find the eigenvalues and eigenvectors of M, we use the function EIGENVECTORS:

(C7) EIGENVECTORS(M);

$$
(D7) \; [[[-\frac{\sqrt{4A+5}-1}{2}, \; \frac{\sqrt{4A+5}+1}{2}, \; -1],
$$

$$
[1, 1, 1]], \; [1, -\frac{\sqrt{4A+5}-1}{2A+2}, \; -\frac{\sqrt{4A+5}-1}{2A+2}],
$$

$$
[1, \frac{\sqrt{4A+5}+1}{2A+2}, \; \frac{\sqrt{4A+5}+1}{2A+2}], \; [1, -1, 0]]
$$

In D7, the first triple gives the eigenvalues of M and the next gives their respective multiplicities (here each is unrepeated). The next three triples

give the corresponding eigenvectors of M.   In order to extract from this
expression one of these eigenvectors, we may use the PART function:

(C8) PART(%,2);

$$
(D8) \quad [1, \; - \; \frac{SQRT(4 A + 5) - 1}{2 A + 2}, \; - \; \frac{SQRT(4 A + 5) - 1}{2 A + 2}]
$$

Programming in MACSYMA

So far we have used MACSYMA in the interactive mode, rather like a calculator.
However, for computations which involve a repetitive sequence of commands like
the ones we present in this book, it is better to execute a program.   Here we
present a short sample program to calculate the critical points of a function f
of two variables x and y.   The program cues the user to enter the function f,
then it computes the partial derivatives f_x and f_y, and then it uses the
MACSYMA command SOLVE to obtain solutions to $f_x = f_y = 0$.   The program is
written outside of MACSYMA with an editor, and then loaded into MACSYMA with
the BATCH command.   Here is the program listing:

```
CRITPTS( ):=(
PRINT("PROGRAM TO FIND CRITICAL POINTS"),
F:READ("ENTER F(X,Y)"),
PRINT("F =",F),
EQS:[DIFF(F,X),DIFF(F,Y)],
UNK:[X,Y],
SOLVE(EQS,UNK))$
```

The program (which is actually a function with no argument) is called
CRITPTS.   Each line is a valid MACSYMA command which could be executed from the
keyboard, and which is separated from the next command by a comma.   The partial

derivatives are stored in a variable named EQS, and the unknowns are stored in UNK.

Here is a sample run:

This is EUNICE MACSYMA Beta Test Release 308.2.
[copyright message...]

(c1) BATCH(CRITPTS)$
[program is loaded in...]

(c4) CRITPTS();
PROGRAM TO FIND CRITICAL POINTS
ENTER F(X,Y)
%E^(X^3+Y^2)*(X+Y);

$$F = (y + x)\ \%e^{y^2 + x^3}$$

(d4) [[x = 0.4588955685487001 %i + 0.3589790871086935,

y = 0.4942017368275118 %i − 0.1225787367783657],

[x = 0.3589790871086935 − 0.4588955685487001 %i,

y = − 0.4942017368275118 %i − 0.1225787367783657],

[x = 0.4187542327234816 %i − 0.6923124204420268,

y = 0.455912070111699 − 0.869726269281412 %i],

[x = − 0.4187542327234816 %i − 0.6923124204420268,

y = 0.869726269281412 %i + 0.455912070111699]]

Partial List of MACSYMA functions

See the MACSYMA Reference Manual [26] for more information. Online information about any function is available through the command DESCRIBE(function name).

ALLROOTS(A) finds all the (generally complex) roots of the polynomial equation A, and lists them in NUMERical format (i.e. to 9 significant figures).

APPEND(A,B) appends the list B to the list A, resulting in a single list.

BATCH(A) loads and runs a BATCH program with filename A.

COEFF(A,B,C) gives the coefficient of B raised to the power C in expression A. C may be omitted if it is unity.

CONCAT(A,B) creates the symbol AB.

CONS(A,B) adds A to the list B as its first element.

DEMOIVRE(A) transforms all complex exponentials in A to their trigonometric equivalents.

DENOM(A) gives the denominator of A.

DEPENDS(A,B) declares A to be a function of B. This is useful for writing unevaluated derivatives, as in specifying differential equations.

DESOLVE(A,B) attempts to solve a linear system A of o.d.e.'s for unknowns B using Laplace transforms. See the Manual for details.

DETERMINANT(A) returns the determinant of the square matrix A.

DIFF(A,B1,C1,B2,C2,...,Bn,Cn) gives the mixed partial derivative of A with respect to each Bi, Ci times. For brevity, DIFF(A,B,1) may be represented by DIFF(A,B). 'DIFF(...) represents the unevaluated derivative, useful in specifying a differential equation.

EIGENVALUES(A) returns two lists, the first being the eigenvalues of the square matrix A, and the second being their respective multiplicities.

EIGENVECTORS(A) does everything that EIGENVALUES does, and adds a list of the eigenvectors of A.

ENTERMATRIX(A,B) cues the user to enter an AxB matrix, element by element.

EV(A,B1,B2,...,Bn) evaluates A subject to the conditions Bi. In particular the Bi may be equations, lists of equations (such as that returned by SOLVE), or assignments, in which cases EV "plugs" the Bi into A. The Bi may also be words such as NUMER (in which case the result is returned in numerical format), DETOUT (in which case any matrix inverses in A are performed with the determinant factored out), or DIFF (in which case all differentiations in A are evaluated, i.e. 'DIFF in A is replaced by DIFF). For brevity in a manual command (i.e., not inside a user defined function), the EV may be dropped, shortening the syntax to A,B1,B2,...,Bn.

EXPAND(A) algebraically expands A. In particular multiplication is distributed over addition.

EXPONENTIALIZE(A) transforms all trigonometric functions in A to their complex exponential equivalents.

FACTOR(A) factors A.

FREEOF(A,B) is true if the variable A is not part of the expression B.

GRIND(A) displays a variable or function A in a compact format. When used with WRITEFILE and an editor outside of MACSYMA, it offers a scheme for producing BATCH files which include MACSYMA generated expressions.

IDENT(A) returns an AxA identity matrix.

IMAGPART(A) returns the imaginary part of A.

INTEGRATE(A,B) attempts to find the indefinite integral of A with respect to B.

INTEGRATE(A,B,C,D) attempts to find the definite integral of A with respect to B taken from B=C to B=D. The limits of integration C and D may be taken as INF (positive infinity) or MINF (negative infinity).

INVERT(A) computes the inverse of the square matrix A.

KILL(A) removes the variable A together with all its assignments and properties from the current MACSYMA environment.

LIMIT(A,B,C) gives the limit of expression A as variable B approaches the value C. The latter may be taken as INF or MINF as in INTEGRATE.

LHS(A) gives the left hand side of the equation A.

LOADFILE(A) loads a disk file with filename A from the current default
directory. The disk file must be in the proper format (i.e. created by a SAVE
command).

MAKELIST(A,B,C,D) creates a list of A's (each of which presumably depends on
B), concatenated from B=C to B=D.

MAP(A,B) maps the function A onto the subexpressions of B.

MATRIX(A1,A2,....,An) creates a matrix consisting of the rows Ai, where each row
Ai is a list of m elements, [B1,B2,....,Bm].

NUM(A) gives the numerator of A.

ODE2(A,B,C) attempts to solve the first or second order ordinary differential
equation A for B as a function of C.

PART(A,B1,B2,....,Bn) first takes the B1th part of A, then the B2th part of
that, and so on.

PLAYBACK(A) displays the last A (an integer) labels and their associated
expressions. If A is omitted, all lines are played-back. See the Manual for
other options.

RATSIMP(A) simplifies A and returns a quotient of two polynomials.

REALPART(A) returns the real part of A.

RHS(A) gives the right hand side of the equation A.

SAVE(A,B1,B2,....,Bn) creates a disk file with filename A in the current default directory, of variables, functions, or arrays Bi. The format of the file permits it to be reloaded into MACSYMA using the LOADFILE command. Everything (including labels) may be SAVE'd by taking B1 equal to ALL.

SOLVE(A,B) attempts to solve the algebraic equation A for the unknown B. A list of solution equations is returned. For brevity, if A is an equation of the form C = 0 it may be abbreviated simply by the expression C.

SOLVE([A1,A2,....,An],[B1,B2,....,Bn]) attempts to solve the system of n polynomial equations Ai for the n unknowns Bi. A list of solution equations is returned.

STRINGOUT(A,B1,B2,....,Bn) creates a disk file with filename A in the current default directory, of variables (e.g. labels) Bi. The file is in a text format and is not reloadable into MACSYMA. However the strungout expressions can incorporated into a FORTRAN or BASIC program with a minimum of editing.

SUBST(A,B,C) substitutes A for B in C.

SUM(A,B,C,D) sums expression A as B varies from C to D.

TAYLOR(A,B,C,D) expands A in a Taylor series in B about B=C, up to and including the term (B-C)^D. MACSYMA also supports Taylor expansions in more than one independent variable; see the Manual for details.

TRANSPOSE(A) gives the transpose of the matrix A.

TRIGEXPAND(A) is a trig simplification function which uses the sum-of-angles formulas to simplify the arguments of individual SIN or COS's. E.g. TRIGEXPAND(SIN(X+Y)) gives COS(X) SIN(Y) + SIN(X) COS(Y).

TRIGREDUCE(A) is a trig simplification function which uses trig identities to convert products and powers of SIN and COS into a sum of terms, each of which contains only a single SIN or COS. E.g., TRIGREDUCE(SIN(X)^2) gives (1 - COS(2X))/2.

TRIGSIMP(A) is a trig simplification function which replaces TAN, SEC, etc., by their SIN and COS equivalents. It also uses the identity SIN()^2 + COS()^2 = 1.

Exercises

1. Integrate $(\log(x))^{20}$, then check your result by differentiating it.

2. Define z to be x + i y (where i is the square root of -1, i.e. %I), and then F to be $\sin(z^2 + \log(z))$. Then find the REALPART of F, call it R, and compute $\frac{\partial^2 R}{\partial x^2} + \frac{\partial^2 R}{\partial y^2}$, which should be zero. Be sure to RATSIMP your answer.

3. Find the equation of the circle passing through the points: (-2,7), (-4,1), (4,-5).

Hint: Look for the solution in the form

$$(x - a)^2 + (y - b)^2 = r^2$$

Use the given points to obtain 3 equations of the above form, and then SOLVE for the 3 unknowns a,b,r.

4. Check the matrix identity $(A\ B)^{-1} = B^{-1}\ A^{-1}$ by computing both sides when

$$A = \begin{bmatrix} 1 & 2 & 3 \\ 4 & 5 & 6 \\ 7 & 8 & K \end{bmatrix} \qquad B = \begin{bmatrix} -2 & 1 & 0 \\ 1 & -2 & 1 \\ 0 & 1 & -2 \end{bmatrix}$$

Use ENTERMATRIX to define A and B, then INVERT both. Remember that matrix multiplication is represented by a period. Be sure to RATSIMP your answer.

5. Find a particular solution to the following fourth order differential equation:

$$y'''' - y''' - y'' - y' - 2y = x^5$$

Hint: Look for a solution in the form of a 5th degree polynomial in x with unknown coefficients. Check your answer by substituting it back into the

REFERENCES

1. Armbruster,D.
   O(2)-Symmetric Bifurcation Theory for Convection Rolls
   to appear in Physica D (1987)

2. Armbruster,D.
   Computer Algebra Programs for Dynamical Systems Theory
   to appear in Proc. Int. Symp. On Pattern Formation, Tuebingen 1986,
   Eds. Dangelmayr,G., Guettinger,W., Springer-Verlag, Heidelberg

3. Arnold,V.I.
   Mathematical Methods of Classical Mechanics
   Springer-Verlag, New York (1978)

4. Auchmuty,J.F.G. and Nicolis,G.
   Bifurcation Analysis of Non-linear Reaction-Diffusion Equations I
   Bull. Math. Biol. 37:323-365 (1975)

5. Bogdanov,R.I.
   Versal Deformations of a Singular Point on the Plane in the Case of Zero
   Eigenvalues
   Functional Analysis and Its Applications, 9:144-145 (1975)

6. Carr,J.
   Applications of Centre Manifold Theory
   Applied Mathematical Sciences, 35
   Springer-Verlag, New York (1981)

7. Cary,J.R.
   Lie Transform Perturbation Theory for Hamiltonian Systems
   Physics Reports 79:129-159 (1981)

8. Cesari,L.
   Asymptotic Behavior and Stability Problems in Ordinary Differential
   Equations
   Ergebnisse der Mathematik und ihrer Grenzgebiete, 16 (Third edition)
   Springer-Verlag, New York (1971)

9. Chakraborty,T.
   Bifurcation Analysis of Two Weakly Coupled van der Pol Oscillators
   Ph.D. thesis, Cornell University (1987)

10. Chakraborty,T. and Rand,R.H.
    The Transition from Phase Locking to Drift in a System of Two Weakly
    Coupled van der Pol Oscillators
    in Transactions of Fourth Army Conference on Applied Mathematics and
    Computing, ARO Report 87-1, pp.1003-1018 (1987)

11. Cushman,R. and Sanders,J.A.
    Nilpotent Normal Forms and Representation Theory of $sl(2,\mathbb{R})$
    Contemporary Math. 56:31-51 (1986)

12. Goldstein,H.
    Classical Mechanics
    Addison-Wesley, Reading (1980, 2nd edition)

13. Golubitsky,M. and Schaeffer,D.G.
    Singularities and Groups in Bifurcation Theory, Vol.I
    Applied Mathematical Sciences, 51
    Springer-Verlag, New York (1985)

14. Guckenheimer,J. and Holmes,P.
    Nonlinear Oscillations, Dynamical Systems and Bifurcations of Vector Fields
    Applied Mathematical Sciences, 42
    Springer-Verlag, New York (1983)

15. Guckenheimer,J. and Knobloch,E.
    Nonlinear Convection in a Rotating Layer: Amplitude Expansions and Normal
    Forms
    Geophys. Astrophys. Fluid Dyn. 23:247-272 (1983)

16. Hale,J.K.
    Ordinary Differential Equations
    Wiley, New York (1969)

17. Holmes,C. and Holmes,P.J.
    Second Order Averaging and Bifurcations to Subharmonics in Duffing's
    Equation
    J.Sound Vib. 78:161-174 (1981)

18. Holmes,C.A. and Rand,R.H.
    Coupled Oscillators as a Model for Parametric Excitation
    Mechanics Research Communications, 8:263-268 (1981)

19. Keith,W.L. and Rand,R.H.
    Dynamics of a System Exhibiting the Global Bifurcation of a Limit Cycle at
    Infinity
    Int.J.Non-Linear Mechanics 20:325-338 (1985)

20. Kevorkian,J. and Cole,J.D.
    Perturbation Methods in Applied Mathematics
    Applied Mathematical Sciences, 34
    Springer-Verlag, New York (1981)

21. Laure,P. and Demay,Y.
    Symbolic Computation and Equations on the Center Manifold
    preprint, Nice (1986)

22. Len,J.
    Nonlinear Parametric Excitation using Lie Transforms and Averaging
    Ph.D. thesis, Cornell University (1987)

23. Len,J. and Rand,R.H.
Nonlinear Parametric Excitation
to appear in Transactions of Fifth Army Conference on Applied Mathematics
and Computing (1987)

24. Lichtenberg,A.J. and Lieberman,M.A.
Regular and Stochastic Motion
Applied Mathematical Sciences, 38
Springer-Verlag, New York (1983)

25. Love,A.E.H.
A Treatise on the Mathematical Theory of Elasicity
Cambridge, University Press (1927, 4th edition)

26. MACSYMA Reference Manual, Version 11 (1986)
prepared by the MACSYMA group of SYMBOLICS, Inc.
11 Cambridge Center, Cambridge, MA 02142

27. Minorsky,N.
Nonlinear Oscillations
Van Nostrand, Princeton (1962)

28. Month,L.A.
On Approximate First Integrals of Hamiltonian Systems with an Application
to Nonlinear Normal Modes in a Two Degree of Freedom Nonlinear Oscillator
Ph.D. thesis, Cornell University (1979)

29. Month,L.A. and Rand,R.H.
An Application of the Poincare Map to the Stability of Nonlinear Normal
Modes
J.Applied Mechanics, 47:645-651 (1980)

30. Month,L.A. and Rand,R.H.
Bifurcation of 4:1 Subharmonics in the Nonlinear Mathieu Equation
Mechanics Research Communications, 9:233-240 (1982)

31. Moon,F.C. and Rand,R.H.
Parametric Stiffness Control of Flexible Structures
in Proceedings of the Workshop on Identification and Control of Flexible
Space Structures, Vol.II,pp.329-342
Jet Propulsion Laboratory Publication 85-29
California Institute of Technology, Pasedena (1985)

32. Nayfeh,A.H.
Perturbation Methods
Wiley, New York (1973)

33. Nicolis,G. and Prigogine,I.
Thermodynamic Theory of Structure, Stability and Fluctuations
Wiley, New York (1977)

34. Normand,C., Pomeau,Y. and Velarde,M.G.
Convective Instability: A Physicist's Approach
Rev. Mod. Phys. 49:581-624 (1977)

35. Rand,R.H.
Computer Algebra in Applied Mathematics: An Introduction to MACSYMA
Research Notes in Mathematics, 94
Pitman Publishing, Boston (1984)

36. Rand,R.H.
    Derivation of the Hopf Bifurcation Formula Using Lindstedt's Perturbation
    Method and MACSYMA
    in Applications of Computer Algebra, R.Pavelle, ed.,293-308
    Kluwer Academic Publishers, Boston (1985)

37. Rand,R.H. and Holmes,P.J.
    Bifurcation of Periodic Motions in Two Weakly Coupled van der Pol
    Oscillators
    Int.J.Non-Linear Mechanics 15:387-399 (1980)

38. Rand,R.H. and Keith,W.L.
    Normal Form and Center Manifold Calculations on MACSYMA
    in Applications of Computer Algebra, R.Pavelle, ed.,309-328
    Kluwer Academic Publishers, Boston (1985)

39. Rand,R.H. and Keith,W.L.
    Determinacy of Degenerate Equilibria with Linear Part x'=y, y'=0 Using
    MACSYMA
    Applied Mathematics and Computation 21:1-19 (1987)

40. Reiss,E.L.
    Column Buckling - An Elementary Example of Bifurcation
    in: Bifurcation Theory and Nonlinear Eigenvalue Problems
    Eds. Keller,J.B. and Antmann,S.
    Benjamin, New York (1969)

41. Sanders,J.A. and Verhulst,F.
    Averaging Methods in Nonlinear Dynamical Systems
    Applied Mathematical Sciences 59
    Springer-Verlag, New York (1985)

42. Sattinger,D.H.
    Topics in Stability and Bifurcation Theory
    Lecture Notes in Mathematics 309
    Springer-Verlag, Berlin (1973)

43. Sattinger,D.H.
    Group Theoretic Methods in Bifurcation Theory
    Lecture Notes in Mathematics 762
    Springer-Verlag, Berlin (1979)

44. Stoker,J.J.
    Nonlinear Vibrations
    Wiley Interscience, New York (1950)

45. Takens,F.
    Singularities of Vector Fields
    Publ.Math.Inst.Hautes Etudes Sci. 43:47-100 (1974)

46. SCRATCHPAD II Newsletter
    Vol.1, No.1, Sept.1985
    Sutor,R.S., Ed.
    Computer Algebra Group
    IBM Watson Research Center
    Box 218 Yorktown Heights, NY 10598

INDEX

Programs are in CAPITAL letters, and their listings are <u>underlined</u>.

Applied Mathematical Sciences

cont. from page ii